THE PRIME NUMBER CONSPIRACY

THE PRIME NUMBER CONSPIRACY

The Biggest Ideas in Math from *Quanta*

edited by Thomas Lin

Quantamagazine

The MIT Press

This book was set in Stone Serif Medium by Westchester Publishing Services. Printed and bound in the United States of America.

Library of Congress Cataloging-in-Publication Data

Names: Lin, Thomas (Journalist), editor.
Title: The prime number conspiracy : the biggest ideas in math from
    Quanta / edited by Thomas Lin ; foreword by James Gleick.
Other titles: Quanta math stories
Description: Cambridge, MA : The MIT Press, [2018] | Includes bibliographical
    references and index.
Identifiers: LCCN 2018013314 | ISBN 9780262536356 (pbk. : alk. paper)
Subjects: LCSH: Mathematics--Popular works.
Classification: LCC QA93 .P75 2018 | DDC 510--dc23 LC record available
at https://lccn.loc.gov/2018013314

10 9 8 7 6 5 4 3 2 1

Men fear thought as they fear nothing else on earth—more than ruin, more even than death. Thought is subversive and revolutionary, destructive and terrible; thought is merciless to privilege, established institutions, and comfortable habits; thought is anarchic and lawless, indifferent to authority, careless of the well-tried wisdom of the ages. Thought looks into the pit of hell and is not afraid. It sees man, a feeble speck, surrounded by unfathomable depths of silence; yet it bears itself proudly, as unmoved as if it were lord of the universe. Thought is great and swift and free, the light of the world, and the chief glory of man.

—Bertrand Russell, *Why Men Fight*

CONTENTS

III   HOW ARE SURPRISING PROOFS DISCOVERED?

IV   HOW DO THE BEST MATHEMATICAL MINDS WORK?

V WHAT CAN OR CAN'T COMPUTERS DO?

VI WHAT IS INFINITY?

VII   IS MATHEMATICS GOOD FOR YOU?

FOREWORD

James Gleick

I t is sometimes said that writing about music is like dancing about architecture—a category mistake. If so, where does that leave writing about mathematics? The writer has only words, and the mathematician inhabits a different place entirely.

Like music, mathematics draws on deep wells of creativity and inspiration, and it can be dark down there. Even the best mathematicians have trouble explaining their own uncanny mental lives. This creates a challenge for the poor journalist. Long ago, I asked Benoit Mandelbrot, the founder of fractal geometry, to describe the source of his intuition about these fantastic shapes and the peculiar methods he had invented. (By "intuition," mathematicians don't mean clairvoyance but rather a sense of what is correct.) He explained it as merely an exercise of the will: "Intuition is not something that is given. I've trained my intuition to accept as obvious shapes which were initially rejected as absurd, and I find everyone else can do the same." In one of the fine profiles and interviews you'll find in this book, Siobhan Roberts presses the great Michael Atiyah, now 89 years old, to describe how inspiration arrives, and he at least tries: "The idea floats in from heaven knows where. It floats around in the sky; you look at it and admire its colors. It's just there. And then at some stage, when you try to freeze it, put it into a solid frame, or make it face reality, then it vanishes; it's gone."

Why should they have to explain, anyway? Let mere mortals do that for them.

As you will see repeatedly in these pages, inspiration strikes willy-nilly. Petr Šeba was at a bus stop in Cuernavaca, Mexico, watching drivers pay cash for slips of paper that recorded the departure times of the buses ahead. It made him think of quantum chaotic systems. Yitang Zhang was in a friend's backyard in Colorado, waiting to leave for a concert, when "the solution suddenly came to him"—a way forward to proving a landmark theorem of number theory. A retired German statistician was brushing his teeth when he saw the key to solving the decades-old Gaussian correlation inequality. (Natalie Wolchover describes this moment as Thomas Royen's "bathroom sink epiphany.") Like so many others, Royen struggles to find words to express the ineffable joy that comes with this kind of discovery. "It is like a kind of grace," he told Wolchover. "We can work for a long time on a problem and suddenly an angel—[which] stands here poetically for the mysteries of our neurons—brings a good idea."

They are just a few of the pioneers we meet here. Maryna Viazovska, a Ukrainian now in Switzerland, packs spheres in eight dimensions while Michaël Rao in France tiles the plane. Martin Hairer remembers exploring the Mandelbrot set when he was a 13-year-old with a Macintosh II. Maryam Mirzakhani wins the Fields Medal for making connections between the geometry of hyperbolic surfaces and physics of dynamical systems. And Artur Avila, another Fields medalist, solves—spoiler alert—the Ten Martini Problem.

Of the arts and sciences, mathematics is among the most ancient and, at the same time, the most modernistic. It can be beautiful and it can be arcane, and its practitioners treasure both qualities. People who write about mathematics and mathematicians learn to embrace these contradictions. In my own early days exploring the subject as a journalist with no mathematical training, I resented the notion that I was, or should be, a "popularizer." The best writing on mathematics does not aim merely to translate or explain. It brings news: bulletins from the frontiers of thought. And it provides us with new ways of seeing the world around us—even the parts that are invisible.

What do you do when you get to the gnarly bits, the intricacies that seem just too difficult to grasp? Whether you're a reader or a reporter, I think the answer's the same: plunge ahead. It is necessary to contemplate the nature of infinity. All right, then! On the one hand, there is not just one infinity; there are many. On the other hand, infinity may not exist; it may not be part of reality; it may be a creature of our (limitless) imagination. If along the way we need to wrap our minds around Ramsey's theorem for pairs, or RT_2^2, so be it. I treasure something David Foster Wallace once wrote: "It is at the higher and apical levels of geometry, topology, analysis, number theory and mathematical logic that the fun and profundity start, when the calculators and contextless formulae fall away and all that's left are pencil and paper and what gets called 'genius,' viz. the particular blend of reason and ecstatic creativity that characterizes what is best about the human mind."

While I'm quoting Wallace, here is an apt footnote he attached to a review in the journal *Science*. He worries that anyone writing about mathematics must bear in mind the wide variability in readers' prior knowledge. He quotes G. H. Hardy's famous 1940 essay, *A Mathematician's Apology*: "On the one hand my examples must be very simple, and intelligible to a reader who has no specialized mathematical knowledge...And on the other hand my examples should be drawn from 'pukka' mathematics, the mathematics of the working professional mathematician." Even writing for *Science*, Wallace realizes that he doesn't know whether he needs to spell out the definition of the Goldbach conjecture: that every even integer greater than 2 can be expressed as the sum of two prime numbers. "As it happens," Wallace's footnote concludes, "this reviewer is not certain whether it's necessary or not, and the fact that these lines have not been deleted by *Science*'s editors (i.e., that you are reading them at all) may indicate that the editors are not totally sure either." Writing about mathematics can get very meta.

Some of the greatest problems are also some of the oldest. The highest peaks of mathematics remain in sight of what we learned as children. Several of the breakthroughs reported in this book involve prime numbers. You know what prime numbers are, but it's customary to explain—"numbers divisible only by 1 and themselves," or some such. Many proofs concerning prime numbers fill the annals of history, one being that there are infinitely many of them. This was proved by Euclid, whose dates of birth and death have been lost in the mists of time; who lacked internet access and electricity and, for that matter, books; who had only papyrus to write on and scarcely anyone to appreciate his discoveries. Euclid did not know that he lived on a planet orbiting the sun, but he knew that prime numbers continue to infinity, and he proved it.

Where proofs have not yet been found, we have conjectures, such as Goldbach's, that may someday grow up to be proofs. If you can talk about primes, you can talk about the gaps between primes. Often the gap is just 2, as with 11 and 13. They are twin primes. How many of these pairs are there? No one knows, but the twin primes conjecture says they are infinite. By the way—if mathematicians were physicists, they would say these conjectures are true. The weight of evidence is enormous. All their experience says so. Mathematicians peer deeper and deeper into their universe, computers serving as their telescopes and spaceships. As they calculate larger and larger numbers, both the Goldbach and the twin primes conjectures continue to hold up. But that's not good enough for a mathematician. A proof is needed to provide certainty. And, in some sense, a proof is needed to explain *why*.

In reporting on new progress in the understanding of primes and prime pairs, Erica Klarreich tells us about sieves, a method for finding primes by filtering out the nonprimes. Sieves go back to Eratosthenes, but the algorithms are improving. Sieves lead to combs, which have teeth. All these words have powerful new meanings assigned by specialists: Groups. Motives. Weights. Graphs. Matroids. Doughnuts.

One well-known feature of prime numbers is that there is a fundamental randomness in their distribution. That randomness is punctured, however, by surprising patterns. The interweaving of randomness and regularity—underlying structure amid disorder—is a running theme in this book, as indeed it is throughout modern science. "A mathematician, like a painter or a poet, is a maker of patterns," Hardy wrote in his *Apology*. His insistence on the apparently unscientific quality of beauty as a touchstone reflects an appreciation of that mixture of order and disorder because we humans perceive neither pure randomness nor perfect regularity as beautiful. We appreciate a bit of chaos.

Perhaps there was a time when mathematics was pure and the lesser sciences coped with real-world messiness, but if so, that time is gone. Mathematicians often lead the way in seeing connections between seemingly unrelated physical systems. As Kevin Hartnett reports, the geometry of random walks may apply as well to quantum strings as it does to bacterial colonies. Mathematicians studying "periods of motives" are drawing links between algebraic geometry and the Feynman diagrams favored by particle physicists. This is what the physicist Eugene Wigner used to call "the unreasonable effectiveness of mathematics." Mathematicians burrow into their world of Platonic imagination, and what they find there—rules, structures and patterns—reappear in the natural sciences as if by coincidence.

In his essay, Robbert Dijkgraaf quotes two great scientists four centuries apart, each of whom expressed the relationship between mathematics and the rest of science in terms of learning a language. First, Galileo: "Philosophy is written in this grand book, the universe, which stands continually open to our gaze. But the book cannot be understood unless one first learns to comprehend the language and read the letters in which it is composed. It is written in the language of mathematics." And then Richard Feynman: "To those who do not know mathematics it is difficult to get across a real feeling as to the beauty, the deepest beauty, of nature.... If you want to learn about nature, to appreciate nature, it is necessary to understand the language that she speaks in."

A new language, Dijkgraaf suggests, comes to us from "the bizarre world of quantum theory—where things can seem to be in two places at the same time and are subject to the laws of probability." He proposes that mathematicians can learn from the unreasonable effectiveness of quantum theory. Indeed, mathematics and quantum theory are cross-pollinating in unexpected ways. The mathematician Svitlana Mayboroda is using the "landscape function" to perform quantum simulations of how electrons are localized in disordered materials—a breakthrough that may boost the efficiency of LEDs. Other mathematicians are looking at the evolution of quantum states in quasicrystals. The advantage of the quantum language, as Dijkgraaf sees it, is that its "holistic approach" considers all the possibilities of a system as one ensemble.

"This holistic approach of considering everything at once is very much in the spirit of modern mathematics," he tells us, "where the study of 'categories' of objects focuses much more on the mutual relations than on any specific individual example." Like physicists, when mathematicians discover facts about their universe, they want to connect them to other facts. They want nature to make sense.

INTRODUCTION

Thomas Lin
Quanta Magazine editor in chief

I t's hard to beat a good science or math story.
Take the events of July 4, 2012. That morning, scientists at the world's biggest physics experiment, the Large Hadron Collider near Geneva, Switzerland, made the biggest announcement of their lives. After investing 20 years in the effort to design and construct the LHC, only to suffer a demoralizing malfunction soon after it went online in 2008, they had finally discovered the Higgs boson, a particle without which life and the universe as we know it could not exist. The next year, Peter W. Higgs and François Englert, whose theoretical work in the 1960s predicted the Higgs particle, won the Nobel Prize in Physics.

The documentary film *Particle Fever* chronicled the hopes and dreams of the thousands of researchers behind the Higgs discovery. In one of my favorite scenes, the theorist David Kaplan is shown in 2008 explaining to a packed lecture hall why they built what the experimentalist Monica Dunford calls "a five-story Swiss watch." Unmoved by Kaplan's talk, an economist demands to know what people stand to gain from the multibillion-dollar experiment: "What's the economic return? How do you justify all this?"

"I have no idea," Kaplan answers bluntly. He must get this question all the time. Patiently, he explains that big breakthroughs in basic science "occur at a level where you're not asking, 'What is the economic gain?' You're asking, 'What do we not know, and where can we make progress?'" In their purest forms, science and mathematics are not about engineering practical applications or cashing in on them, though those things often do happen later, sometimes much later. They're about learning something you didn't know before.

"So, what is the LHC good for?" Kaplan asks the economist, setting up the death blow: "It could be nothing, other than just—understanding everything."

As it happens, this book picks up where *Particle Fever* leaves off in telling the story of the quest to understand everything. The renowned theoretical physicist Nima Arkani-Hamed has described such efforts in fundamental physics as "trying to understand, in the simplest possible way, the smallest set of basic principles from which everything else in principle follows." The fewer the assumptions, the approximations, the contortions—or so the thinking goes—the closer we are to the truth. The Higgs boson has been discovered and the Standard Model of particle physics is now complete. The problem is, absent new particles beyond the Standard Model, the universe doesn't make sense. How, then, are we to make sense of it?

In *Alice and Bob Meet the Wall of Fire: The Biggest Ideas in Science from Quanta*, and *The Prime Number Conspiracy: The Biggest Ideas in Math from Quanta*, we join some of the greatest scientific and mathematical minds as they test the limits of human knowledge. The stories presented in these two companion volumes reveal efforts over the past five years or so to untangle the mysteries of the universe—its origins and basic laws, its contents big and small, its profoundly complex living inhabitants—and to unlock the universal language of nature. They penetrate the big questions and uncover some of the best ideas and theories for understanding our physical, biotic and logical worlds. Meanwhile, they illuminate the essential issues under debate as well as the obstacles hindering further progress.

In selecting and editing *Quanta Magazine* articles for these volumes, I tried to venture beyond the usual mixtape format of "best of" anthologies and greatest hits compilations. Instead, I wanted to send readers on breathtaking intellectual journeys to the leading edge of discovery strapped to the narrative rocket of humanity's never-ending pursuit of knowledge. But what might those excursions actually look like? These nonfiction adventures, it turns out, explore core questions about the essence of prime numbers, whether our universe is "natural," the natures of time and infinity, our strange quantum reality, whether space-time is fundamental or emergent, the insides and outsides of black holes, the origin and evolution of life, what makes us human, the hopes for and limitations of computing, the role of mathematics in science and society and just where these questions are taking us. The stories in these books reveal how cutting-edge research is done—how the productive tension between theory, experiment and mathematical intuition, through triumphs, failures and null results, cuts a path forward.

What is *Quanta*? Albert Einstein called photons "quanta of light." *Quanta Magazine* strives to illuminate the dark recesses in science and mathematics where some of the most provocative and fundamental ideas are cultivated out of public view. Not that anyone is trying to hide them. The work hides in plain sight at highly technical conferences and workshops, on the preprint site arxiv.org and in impenetrable academic journals. These are not easy subjects to understand, even for experts in adjacent fields, so it's not surprising that only Higgs-level discoveries are widely covered by the popular press.

The story of *Quanta* began in 2012, just weeks after the Higgs announcement. With the news industry still reeling from the 2008 financial crisis and secular declines in print advertising, I had the not-so-brilliant idea to start a science magazine. The magazine I envisioned would apply the best editorial standards of publications like the *New York Times* and the *New Yorker*, but its coverage would differ radically from that of existing news outlets. For one thing, it wouldn't report on anything you might actually find useful. This magazine would not publish health or medical news or breathless coverage of the latest technological breakthroughs. There would be no advice on which foods or vitamins to consume or avoid, which exercises to wedge into your day, which gadgets are must-buys. No stories about crumbling infrastructure or awesome feats of engineering. It wouldn't even keep you updated about the latest NASA mission, exoplanet find or SpaceX rocket launch. There's nothing wrong with any of this, of course. When accurately reported, deftly written and carefully fact checked, it's "news you can use." But I had other ideas. I wanted a science magazine that helps us achieve escape velocity beyond our own small worlds but is otherwise useless in the way the LHC is useless. This useless magazine became *Quanta*.

My colleagues and I also treat our readers differently. We don't protect them from the central concepts or from the process of how new ideas come to be. Indeed, the ridiculously difficult science and math problems and the manner in which an individual or collaboration makes progress serve as the very conflicts and resolutions that drive *Quanta* narratives. We avoid jargon, but we don't protect readers from the science itself. We trust readers, whether they have a science background or not, to be intellectually curious enough to want to know more, so we give you more.

Like the magazine, this book is for anyone who wants to understand how nature works, what the universe is made of, and how life got its start and evolved into its myriad forms. It's for curiosity seekers who want a front-row seat for breakthroughs to the biggest mathematical puzzles and whose idea of fun is to witness the expansion of our mathematical universe.

If I may offer an adaptation of Shel Silverstein's famous lyric invitation (my sincere apologies to the late Mr. Silverstein):

If you are a dreamer, come in,
If you are a dreamer, a thinker, a curiosity seeker,
A theorizer, an experimenter, a *mathematiker*...
If you're a tinkerer, come fill my beaker
For we have some mind-bendin' puzzles to examine.
Come in!
Come in!

WHAT'S SO SPECIAL ABOUT PRIME NUMBERS?

UNHERALDED MATHEMATICIAN BRIDGES THE PRIME GAP

Erica Klarreich

O n April 17, 2013, a paper arrived in the inbox of *Annals of Mathematics*, one of the discipline's preeminent journals. Written by a mathematician virtually unknown to the experts in his field—a 50-something lecturer at the University of New Hampshire named Yitang Zhang—the paper claimed to have taken a huge step forward in understanding one of mathematics' oldest problems, the twin primes conjecture.

Editors of prominent mathematics journals are used to fielding grandiose claims from obscure authors, but this paper was different. Written with crystalline clarity and a total command of the topic's current state of the art, it was evidently a serious piece of work, and the *Annals* editors decided to put it on the fast track.

Just three weeks later—a blink of an eye compared to the usual pace of mathematics journals—Zhang received the referee report on his paper.

"The main results are of the first rank," one of the referees wrote. The author had proved "a landmark theorem in the distribution of prime numbers."

Rumors swept through the mathematics community that a great advance had been made by a researcher no one seemed to know—someone whose talents had been so overlooked after he earned his doctorate in 1991 that he had found it difficult to get an academic job, working for several years as an accountant and even in a Subway sandwich shop.

"Basically, no one knows him," said Andrew Granville, a number theorist at the University of Montreal. "Now, suddenly, he has proved one of the great results in the history of number theory."

Mathematicians at Harvard University hastily arranged for Zhang to present his work to a packed audience there on May 13 that year. As details of his work emerged, it became clear that Zhang achieved his result not via a radically new approach to the problem, but by applying existing methods with great perseverance.

"The big experts in the field had already tried to make this approach work," Granville said. "He's not a known expert, but he succeeded where all the experts had failed."

THE PROBLEM OF PAIRS

Prime numbers—those that have no factors other than 1 and themselves—are the atoms of arithmetic and have fascinated mathematicians since the time of Euclid, who proved more than 2,000 years ago that there are infinitely many of them.

Because prime numbers are fundamentally connected with multiplication, understanding their additive properties can be tricky. Some of the oldest unsolved problems in mathematics concern basic questions about primes and addition, such as the twin primes conjecture, which proposes that there are infinitely many pairs of primes that differ by only 2, and the Goldbach conjecture, which proposes that every even number is the sum of two primes. (By an astonishing coincidence, a weaker version of this latter question was settled in a paper posted online by Harald Helfgott of École Normale Supérieure in Paris while Zhang was delivering his Harvard lecture.[1])

Prime numbers are abundant at the beginning of the number line, but they grow much sparser among large numbers. Of the first 10 numbers, for example, 40 percent are prime—2, 3, 5 and 7—but among 10-digit numbers, only about 4 percent are prime. For over a century, mathematicians have understood how the primes taper off on average: Among large numbers, the expected gap between prime numbers is approximately 2.3 times the number of digits; so, for example, among 100-digit numbers, the expected gap between primes is about 230.

But that's just on average. Primes are often much closer together than the average predicts, or much farther apart. In particular, "twin" primes often crop up—pairs such as 3 and 5, or 11 and 13, that differ by only 2. And while such pairs get rarer among larger numbers, twin primes never seem to disappear completely (the largest pair discovered so far is $2,996,863,034,895 \times 2^{1,290,000} \pm 1$).

For hundreds of years, mathematicians have speculated that there are infinitely many twin prime pairs. In 1849, the French mathematician Alphonse de Polignac extended this conjecture to the idea that there should be infinitely many prime pairs for any possible finite gap, not just 2.

Since that time, the intrinsic appeal of these conjectures has given them the status of a mathematical holy grail, even though they have no known applications. But despite many efforts at proving them, mathematicians weren't able to rule out the possibility that the gaps between primes grow and grow, eventually exceeding any particular bound.

Now Zhang has broken through this barrier. His paper shows that there is some number N smaller than 70 million such that there are infinitely many pairs of primes that differ by N. No matter how far you go into the deserts of the truly gargantuan prime numbers—no matter how sparse the primes become—you will keep finding prime pairs that differ by less than 70 million.

The result is "astounding," said Daniel Goldston, a number theorist at San Jose State University. "It's one of those problems you weren't sure people would ever be able to solve."

A PRIME SIEVE

The seeds of Zhang's result lie in a paper from 13 years ago that number theorists refer to as GPY, after its three authors—Goldston, János Pintz of the Alfréd Rényi Institute of Mathematics in Budapest, and Cem Yıldırım of Boğaziçi University in Istanbul.[2] That paper came tantalizingly close but was ultimately unable to prove that there are infinitely many pairs of primes with some finite gap.

Instead, it showed that there will always be pairs of primes much closer together than the average spacing predicts. More precisely, GPY showed that for any fraction you choose, no matter how tiny, there will always be a pair of primes closer together than that fraction of the average gap, if you go out far enough along the number line. But the researchers couldn't prove that the gaps between these prime pairs are always less than some particular finite number.

GPY uses a method called "sieving" to filter out pairs of primes that are closer together than average. Sieves have long been used in the study of prime numbers, starting with the 2,000-year-old Sieve of Eratosthenes, a technique for finding prime numbers.

To use the Sieve of Eratosthenes to find, say, all the primes up to 100, start with the number two, and cross out any higher number on the list that is divisible by two. Next move on to three, and cross out all the numbers divisible by three. Four is already crossed out, so you move on to five, and cross out all the numbers divisible by five, and so on. The numbers that survive this crossing-out process are the primes.

The Sieve of Eratosthenes works perfectly to identify primes, but it is too cumbersome and inefficient to be used to answer theoretical questions.

Over the past century, number theorists have developed a collection of methods that provide useful approximate answers to such questions.

"The Sieve of Eratosthenes does too good a job," Goldston said. "Modern sieve methods give up on trying to sieve perfectly."

GPY developed a sieve that filters out lists of numbers that are plausible candidates for having prime pairs in them. To get from there to actual prime pairs, the researchers combined their sieving tool with a function whose effectiveness is based on a parameter called the level of distribution that measures how quickly the prime numbers start to display certain regularities.

The level of distribution is known to be at least 1/2.[3] This is exactly the right value to prove the GPY result, but it falls just short of proving that there are always pairs of primes with a bounded gap. The sieve in GPY could establish that result, the researchers showed, but only if the level of distribution of the primes could be shown to be more than 1/2. Any amount more would be enough. The theorem in GPY "would appear to be within a hair's breadth of obtaining this result," the researchers wrote.

But the more researchers tried to overcome this obstacle, the thicker the hair seemed to become. During the late 1980s, three researchers—Enrico Bombieri, a Fields medalist at the Institute for Advanced Study in Princeton, John Friedlander of the University of Toronto, and Henryk Iwaniec of Rutgers University—had developed a way to tweak the definition of the level of distribution to bring the value of this adjusted parameter up to 4/7.[4] After the GPY paper was circulated in 2005, researchers worked feverishly to incorporate this tweaked level of distribution into GPY's sieving framework, but to no avail.

"The big experts in the area tried and failed," Granville said. "I personally didn't think anyone was going to be able to do it anytime soon."

CLOSING THE GAP

Meanwhile, Zhang was working in solitude to try to bridge the gap between the GPY result and the bounded prime gaps conjecture. A Chinese immigrant who received his doctorate from Purdue University, he had always been interested in number theory, even though it wasn't the subject of his dissertation. During the difficult years in which he was unable to get an academic job, he continued to follow developments in the field.

"There are a lot of chances in your career, but the important thing is to keep thinking," he said.

Zhang read the GPY paper, and in particular the sentence referring to the hair's breadth between GPY and bounded prime gaps. "That sentence impressed me so much," he said.

Without communicating with the field's experts, Zhang started thinking about the problem. After three years, however, he had made no progress. "I was so tired," he said.

To take a break, Zhang visited a friend in Colorado during the summer of 2012. There, on July 3, during a half-hour lull in his friend's backyard before leaving for a concert, the solution suddenly came to him. "I immediately realized that it would work," he said.

Zhang's idea was to use not the GPY sieve but a modified version of it, in which the sieve filters not by every number, but only by numbers that have no large prime factors.

"His sieve doesn't do as good a job because you're not using everything you can sieve with," Goldston said. "But it turns out that while it's a little less effective, it gives him the flexibility that allows the argument to work."

While the new sieve allowed Zhang to prove that there are infinitely many prime pairs closer together than 70 million, it is unlikely that his methods can be pushed as far as the twin primes conjecture, Goldston said. Even with the strongest possible assumptions about the value of the level of distribution, he said, the best result likely to emerge from the GPY method would be that there are infinitely many prime pairs that differ by 16 or less.

But Granville said that mathematicians shouldn't prematurely rule out the possibility of reaching the twin primes conjecture by these methods.

"This work is a game changer, and sometimes after a new proof, what had previously appeared to be much harder turns out to be just a tiny extension," he said. "For now, we need to study the paper and see what's what."

It took Zhang several months to work through all the details, but the resulting paper is a model of clear exposition, Granville said. "He nailed down every detail so no one will doubt him. There's no waffling."

Once Zhang received the referee report, events unfolded with dizzying speed. Invitations to speak on his work poured in. "I think people are pretty thrilled that someone out of nowhere did this," Granville said.

For Zhang, who calls himself shy, the glare of the spotlight has been somewhat uncomfortable. "I said, 'Why is this so quick?'" he said. "It was confusing, sometimes."

Zhang was not shy, though, during his Harvard talk, which attendees praised for its clarity. "When I'm giving a talk and concentrating on the math, I forget my shyness," he said.

Zhang said he feels no resentment about the relative obscurity of his career thus far. "My mind is very peaceful. I don't care so much about the money, or the honor," he said. "I like to be very quiet and keep working by myself."

TOGETHER AND ALONE, CLOSING THE PRIME GAP

Erica Klarreich

I n the months after Yitang Zhang settled a long-standing open question about prime numbers in 2013—showing that even though primes get increasingly rare as you go further out along the number line, you will never stop finding pairs of primes separated by at most 70 million—he found himself caught up in a whirlwind of activity and excitement. He lectured on his work at many of the nation's preeminent universities, received job offers from top institutions in China and Taiwan and a visiting position at the Institute for Advanced Study in Princeton, New Jersey, and was promoted to full professor at the University of New Hampshire.

Meanwhile, Zhang's work raised a question: Why 70 million? There is nothing magical about that number—it served Zhang's purposes and simplified his proof. Other mathematicians quickly realized that it should be possible to push this separation bound quite a bit lower, although not all the way down to two.

By the end of May 2013, mathematicians had uncovered simple tweaks to Zhang's argument that brought the bound below 60 million. A blog post by Scott Morrison of the Australian National University in Canberra ignited a firestorm of activity, as mathematicians vied to improve on this number, setting one record after another. By June 4 that year, Terence Tao of the University of California, Los Angeles, a winner of the prestigious Fields Medal had created an open, online "Polymath project" to improve the bound that attracted dozens of participants.

For weeks, the project moved forward at a breathless pace. "At times, the bound was going down every thirty minutes," Tao recalled. By July 27, 2013, the team had succeeded in reducing the proven bound on prime gaps from 70 million to 4,680.

Then, a preprint posted to arxiv.org that November by James Maynard, who was then a postdoctoral researcher working on his own at the University of Montreal, upped the ante.[1] Just months after Zhang announced his result, Maynard presented an independent proof that pushed the gap down to 600. Another Polymath project combined the collaboration's techniques with Maynard's approach to push this bound down to 246.

"The community is very excited by this new progress," Tao said.

Maynard's approach applies not just to pairs of primes, but to triples, quadruples and larger collections of primes. He showed that you can find bounded clusters of any chosen number of primes infinitely often as you go out along the number line. (Tao said he independently arrived at this result at about the same time as Maynard.)

Zhang's work and, to a lesser degree, Maynard's fits the archetype of the solitary mathematical genius, working for years in the proverbial garret until he is ready to dazzle the world with a great discovery. The Polymath project couldn't be more different—fast and furious, massively collaborative, fueled by the instant gratification of setting a new world record.

For Zhang, working alone and nearly obsessively on a single hard problem brought a huge payoff. Would he recommend that approach to other mathematicians? "It's hard to say," he said. "I choose my own way, but it's only my way."

Tao actively discourages young mathematicians from heading down such a path, which he has called "a particularly dangerous occupational hazard" that has seldom worked well, except for established mathematicians with a secure career and a proven track record. However, he said in an interview, the solitary and collaborative approaches each have something to offer mathematics.

"It's important to have people who are willing to work in isolation and buck the conventional wisdom," Tao said. Polymath, by contrast, is "entirely groupthink." Not every math problem would lend itself to such collaboration, but this one did.

COMBING THE NUMBER LINE

Zhang proved his result by going fishing for prime numbers using a mathematical tool called a k-tuple, which you can visualize as a comb with some of its teeth snapped off. If you position such a comb along the number line starting at any chosen spot, the remaining teeth will point to some collection of numbers.

Zhang focused on snapped combs whose remaining teeth satisfy a divisibility property called "admissibility." He showed that if you go fishing for primes using any admissible comb with at least 3.5 million teeth, there are infinitely many positions along the number line where the comb will catch at least two prime numbers. Next, he showed how to make an admissible comb with at least 3.5 million remaining teeth by starting with a 70-million-tooth comb and snapping off all but its prime teeth. Such a comb must catch two primes again and again, he concluded, and the primes it catches are separated by at most 70 million.

The finding is "a fantastic breakthrough," said Andrew Granville of the University of Montreal. "It's a historic result."

Zhang's work involved three separate steps, each of which offered potential room for improvement on his 70 million bound. First, Zhang invoked some very deep mathematics to figure out where prime fish are likely to be hiding. Next, he used this result to figure out how many teeth his comb would need in order to guarantee that it would catch at least two prime fish infinitely often. Finally, he calculated how large a comb he had to start with so that enough teeth would be left after it had been snapped down to admissibility.

The fact that these three steps could be separated made improving Zhang's bound an ideal project for a crowd-sourced collaboration, Tao said. "His proof is very modular, so we could parallelize the project, and people with different skills squeezed out what improvements they could."

The Polymath project quickly attracted people with the right skills, perhaps more efficiently than if the project had been organized from the top down. "A Polymath project brings together people who wouldn't have thought of coming together," Tao said.

PRIME FISHING GROUNDS

Of Zhang's three steps, the first to admit improvement was the last one, in which he found an admissible comb with at least 3.5 million teeth. Zhang had shown that a comb of length 70 million would do the trick, but he hadn't tried particularly hard to make his comb as small as possible. There was plenty of room for improvement, and researchers who were good at computational mathematics soon started a friendly race to find small admissible combs with a given number of teeth.

Andrew Sutherland, of the Massachusetts Institute of Technology, quickly became a sort of de facto admissible-comb czar. Sutherland, who focuses on computational number theory, had been traveling during Zhang's announcement and hadn't paid particular attention to it. But when he checked in at a Chicago hotel and mentioned to the clerk that he was there for a mathematics conference, the clerk replied, "Wow, 70 million, huh?"

"I was floored that he knew about it," Sutherland said. He soon discovered that there was plenty of scope for someone with his computational skills to help improve Zhang's bound. "I had lots of plans for the summer, but they went by the wayside."

For the mathematicians working on this step, the ground kept shifting underfoot. Their task changed every time the mathematicians working on the other two steps managed to reduce the number of teeth the comb would require. "The rules of the game were changing on a day-to-day basis," Sutherland said. "While I was sleeping, people in Europe would post new bounds. Sometimes, I would run downstairs at 2 a.m. with an idea to post."

The team eventually came up with the Polymath project's record-holder—a 632-tooth comb whose width is 4,680—using a genetic algorithm that "mates" admissible combs with each other to produce new, potentially better combs.[2]

Maynard's finding, which involves a 105-tooth comb whose width is 600, rendered these giant computations obsolete. But the team's effort was not a wasted one: Finding small admissible combs plays a part in many number theory problems, Sutherland said. In particular, the team's computational tools will likely prove useful when it comes to refining Maynard's results about triples, quadruples and larger collections of primes, Maynard said.

The Polymath researchers focusing on step two of Zhang's proof looked for places to position the comb along the number line that had the greatest likelihood of catching pairs of primes, to figure out the number of teeth required. Prime numbers become very sparse as you go out along the number line, so if you just plunk your comb down somewhere randomly, you probably won't catch any primes, let alone two. Finding the richest fishing grounds for prime numbers ended up being a problem in "calculus of variations," a generalization of calculus.

This step involved perhaps some of the least novel developments in the project, and the ones that were most directly superseded by Maynard's work. At the time, though, this advance was one of the most fruitful ones. When the team filled in this piece of the puzzle on June 5, 2013, the bound on prime gaps dropped from about 4.6 million to 389,922.

The researchers focusing on step one of Zhang's proof, which deals with how prime numbers are distributed, had perhaps the hardest job. Mathematicians have been familiar with a collection of distribution laws for primes for more than a century. One such law says that if you divide all prime numbers by the number three, half of the primes will produce a remainder of one and half will produce a remainder of two. This kind of law is exactly what's needed to figure out whether an admissible comb is likely to find pairs of primes or miss them, since it suggests that "[prime] fish can't all hide behind the same rock, but are spread out everywhere," Sutherland said. But to use such distribution laws in his proof, Zhang—and, later, the Polymath project—had to grapple with some of the deepest mathematics around: a collection of theorems from the 1970s by Pierre Deligne, now an emeritus professor at the Institute for Advanced Study, concerning when certain error terms are likely to cancel each other out in gigantic sums. Morrison described Deligne's work as "a big and terrifying piece of 20th-century mathematics."

"We were very fortunate that several of the participants were well-versed in the difficult machinery that Deligne developed," Tao said. "I myself did not know much about this area until this project."

The project didn't just figure out how to refine this part of the proof to improve the bound. It also came up with an alternative approach that eliminates the need for Deligne's theorems entirely, although at some cost to the bound: Without Deligne's theorems, the best bound the project has come up with is 14,950.

This simplification of the proof is, if anything, more exciting to mathematicians than the final number the project came up with, since mathematicians care not only about whether a proof is correct but also about how much new insight it gives them.

"What we're in the market for are ideas," said Granville.

As the Polymath project progressed, Zhang himself was conspicuously, though perhaps not surprisingly, absent. He has not followed the project closely, he said. "I didn't contact them at all. I prefer to keep quiet and alone. It gives me the opportunity to concentrate."

Also absent, though less conspicuously, was Maynard. As the Polymath participants worked feverishly to improve the bound between prime pairs, Maynard was working on his own to develop a different approach—one foreshadowed by a forgotten paper that was written, and then retracted, ten years ago.

A SECRET WEAPON

Zhang's work was grounded in a 2005 paper known as GPY, after its authors, Daniel Goldston, János Pintz and Cem Yıldırım.[3] The GPY paper developed a scoring system to gauge how close a given number is to being prime. Even numbers get a very low score, odd numbers divisible by 3 are only slightly higher and so on. Such scoring formulas, called sieves, can also be used to score the collection of numbers an admissible comb points to, and they are a crucial tool when it comes to figuring out where to place the comb on the number line so that it has a good chance of catching prime fish. Constructing an effective sieve is something of an art: The formula must provide good estimates of different numbers' prime potential, but it must also be simple enough to analyze.

Two years before GPY was published, two of its authors, Goldston and Yıldırım, had circulated a paper describing what they asserted was a powerful scoring method. Within months, however, mathematicians discovered a flaw in that paper. Once Goldston, Yıldırım and Pintz adjusted the formula to repair this flaw, most mathematicians turned their focus to this adjusted scoring system, the GPY version, and didn't consider whether there might be even better ways to tweak the original, flawed formula.

"Those of us looking at GPY thought we had the bases covered, and it didn't cross our mind to go back and redo the earlier analysis," said Granville, who was Maynard's postdoctoral adviser.

In 2012, however, Maynard decided to go back and take a second look at the earlier paper. A newly minted Ph.D. who had studied sieving theory, he spotted a new way to adjust the paper's scoring system. GPY's approach to scoring an admissible comb had been to multiply together all the numbers the comb pointed to and then score the product in one fell swoop. Maynard figured out a way to score each number separately, thereby deriving much more nuanced information from the scoring system.

Maynard's sieving method "turns out to be surprisingly easy," Granville said. "It's the sort of thing where people like me slap their foreheads and say, 'We could have done this seven years ago if we hadn't been so sure we couldn't do it!'"

With this refined scoring system, Maynard was able to bring the prime gap down to 600 and also prove a corresponding result about bounded gaps between larger collections of primes.

The fact that Zhang and Maynard managed, within months of each other, to prove that prime gaps are bounded is "a complete coincidence," Maynard said. "I found Zhang's result very exciting when I heard about it."

Tao was similarly philosophical about Maynard's scoop of the Polymath project's headline number. "You expect the record to be beaten—that's progress," he said.

It is likely, Tao and Maynard said, that Maynard's sieve can be combined with the deep technical work by Zhang and the Polymath project about the distribution of primes to bring the prime gap even lower.

This time, Maynard joined in. "I'm looking forward to trying to get the bound as small as possible," he said.

It remains to be seen how much more can be wrung out of Zhang's and Maynard's methods. Prior to Maynard's work, the best-case scenario seemed to be that the bound on prime gaps could be pushed down to 16, the theoretical limit of the GPY approach. Maynard's refinements push this theoretical limit down to 12. Conceivably, Maynard said, someone with a clever sieve idea could push this limit as low as 6. But it's unlikely, he said, that anyone could use these ideas to get all the way down to a prime gap of 2 to prove the twin primes conjecture.

"I feel that we still need some very large conceptual breakthrough to handle the twin primes case," Maynard said.

Tao, Maynard and the Polymath participants may eventually get an influx of new ideas from Zhang himself. It has taken the jet-setting mathematician a while to master the art of thinking about mathematics on airplanes, but he has now started working on a new problem, about which he declined to say more than that it is "important." While he isn't currently working on the twin primes problem, he said, he has a "secret weapon" in reserve—a technique to reduce the bound that he developed before his result went public. He omitted this technique from his paper because it is so technical and difficult, he said.

"It's my own original idea," he said. "It should be a completely new thing."

KAISA MATOMÄKI DREAMS OF PRIMES

Kevin Hartnett

P rime numbers are the central characters in mathematics, the indivisible elements from which all other numbers are constructed. Around
300 B.C., Euclid proved that there's an infinite number of them. Millen-
nia later, in the late 19th century, mathematicians refined Euclid's result
and proved that the number of prime numbers over any very large interval
between 1 and some number, x, is roughly $x/\log(x)$.

This estimate, called the prime number theorem, has only been proved
to hold when x is really big, which raises the question—does it hold over
shorter intervals?

Kaisa Matomäki, 33, grew up in the small town of Nakkila in western
Finland, where she was a math star from an early age. She left home to
attend a boarding school that specialized in math instruction, and as a
senior she won first prize in a national mathematics competition. When
she began serious mathematical research as a graduate student, she was
drawn to prime numbers and, in particular, to questions about how they
behave on smaller scales.

In 1896 mathematicians proved that roughly half of all numbers have
an even number of prime factors and half have an odd number. In 2014
Matomäki, now a professor at the University of Turku in Finland, and her
frequent collaborator, Maksym Radziwill of McGill University, proved that
this statement also holds when you look at prime factors over short intervals. The methods they developed to accomplish this have been adopted by
other prominent mathematicians, leading to a number of important results
in the study of primes. For these achievements, Matomäki and Radziwill
shared the 2016 SASTRA Ramanujan Prize, one of the most prestigious
awards for young researchers in number theory.

But Matomäki's results have only deepened the mystery surrounding
primes. As she explained to *Quanta Magazine*, mathematicians had long
assumed that a proof about even and odd numbers of prime factors over

short intervals would lead inexorably to a proof about the prime number theorem over short intervals. Yet, when Matomäki achieved a proof of the first question, she found that a proof of the latter one moved even further out of reach—establishing once again that primes won't be easily cornered.

Quanta Magazine caught up with Matomäki in 2017 to ask about the study of primes and the methods behind her breakthroughs. An edited and condensed version of the conversations follows.

Your work has dealt with two prominent related problems about prime numbers. Could you explain them?

One of the most fundamental theorems in analytic number theory is the prime number theorem, which says that the number of primes up to x is about $x/\log(x)$.

The Riemann zeta function is profoundly connected to the distribution of prime numbers.

This is known to be equivalent to the fact that roughly half of the numbers up to x have an even number of prime factors and half of the numbers have an odd number of prime factors. It's not obvious that the two are equivalent, but it's known that they are equivalent because of facts related to the zeros of the Riemann zeta function.

So these two problems have been known to be equivalent for a long time. Where does your work on them begin?

I've been interested in questions about prime numbers and the prime number theorem, and that's really related to this thing on the number of even and odd prime factors. So what Maksym Radziwill and I have studied is the local distribution of this thing. So even if you take a short sample of numbers, then typically about half of them have an even number of prime factors and half of them have an odd number of prime factors. This doesn't only work from 1 to x, it also works in almost all very short segments.

Let's talk about the methods you used to prove something about these very short segments—a way of working with something called multiplicative functions. These are a class of functions in which $f(m \times n)$ is the same thing as $f(m) \times f(n)$. Why are these functions interesting?

One can use them, for instance, to study these numbers that have an even number of prime factors or an odd number of prime factors. There is a multiplicative function that takes the value -1 if n has an odd number of prime factors and the value $+1$ if n has an even number of prime factors. This is sort of the most important example of a multiplicative function.

And you take these multiplicative functions and do a "decomposition" on them. What does that mean?
We take out a small prime factor from the number n. It's a bit difficult to explain over the phone. We are looking at a typical number, n, in our desired interval, and we notice the number n has a small prime factor, and then we consider this small prime factor separately. So we write it as $p \times m$ where p is a small prime factor.

You started this work when you were looking at sign changes of multiplicative functions. These are situations where the sign (positive or negative) of the output of a function can tell you something about the input. Like, for example, a function that outputs a −1 when the input number has an odd number of prime factors and a +1 when the input number has an even number of prime factors. What were some especially significant moments along the way?
Originally, starting in 2013, Maksym and I worked on sign changes of another sequence that's not related to this, but then we started to think about the more general question of sign changes of multiplicative functions. Then we both spent autumn 2014 in Montreal, where we worked intensively on this problem of sign changes of multiplicative functions and got into shorter and shorter intervals for these sign changes, and eventually at the end of the semester we realized we also get this result about half the numbers in very short intervals having an even number of prime factors and half having an odd number of prime factors.

That wasn't the result you were after all along?
No, no, we didn't think about that at all. We were thinking about sign changes of multiplicative functions, then later on we realized there were more interesting applications than this original problem we were thinking about.

What was it like when you realized the work had this implication?
Of course it was very nice. It doesn't really tell you anything about the primes in the end; it only tells you about even and odd numbers of prime factors. But it is related to the primes, and of course we were very happy to realize that we had more applications, but it was a gradual process to get there. There was no single moment.

You use sieve methods in a lot of your work, though not for this prime number result we've been talking about. "Sieve" is such an evocative name for a math technique. Can you try to capture for people what it's about?

It's meant to sieve some numbers. If you want to get only primes you use some sort of sieve which only lets primes go, and all the composite numbers are held in the sieve. But it's pretty technical to actually do it.

What kinds of things in math act as a sieve?
Essentially what one does is one cooks up a function, which is, say, positive with primes and zero or negative with other numbers.

So you have a function and if the input is a prime, it outputs a positive value?
Yeah, and if the input is not a prime it outputs a zero or a negative value. Instead of primes you study this function, and if you can show this function is positive for some set, say, then you know there are some primes. The point is that the function is easier to handle than the primes themselves.

How long have sieving methods been around?
The first sieve is the sieve of Eratosthenes, an ancient Greek, so very old. He noticed that if you want to generate all the primes, it's enough to make a big table of numbers and cross out all the multiples of 2, then 3, then 5 and so on. You end up with only primes on your table.

When you and Radziwill won the 2016 SASTRA Ramanujan Prize, the citation noted that the methods you've developed have "revolutionized" number theory. What have been some of the most exciting consequences of the work?
We got these averages of multiplicative functions from very short intervals, and it has turned out to be very useful for other things. Terry Tao has worked on problems including the Erdős Discrepancy Problem using our work.

As a result of the discoveries you've made, do you have a different perspective on prime numbers than you used to?
I guess the surprising thing about the primes is that over short intervals, the fact that we can show that half the numbers have an even number of prime factors and half have an odd number doesn't tell us anything about the number of primes.

Could you explain that?
Over long intervals, the prime number theorem is equivalent to half the numbers having an even number of prime factors and half having an odd number of prime factors. But in the very short intervals we are considering, it turns out it's not equivalent.

So it's surprising that over short intervals the two theorems are not equivalent?

What's surprising is that we are able to do this thing about the even number of prime factors and odd number of prime factors, but we are not able to do the prime [number theorem]. It was thought that they are morally equivalent, having the same difficulty, but it turns out they are not.

Was there a moment where you thought this work would give you the prime number theorem over short intervals?

I think we were never optimistic about it. We've been trying it, but we haven't been optimistic. We are still trying it, but it seems that this approach doesn't really work for it, so we have to invent something new.

Is that disappointing?

Certainly it's disappointing. I've been dreaming of doing primes in very short intervals for a long time and still would love to do it.

MATHEMATICIANS DISCOVER PRIME CONSPIRACY

Erica Klarreich

Two mathematicians have uncovered a simple, previously unnoticed property of prime numbers: It seems they have decided preferences about the final digits of the primes that immediately follow them.

Among the first billion prime numbers, for instance, a prime ending in 9 is almost 65 percent more likely to be followed by a prime ending in 1 than another prime ending in 9. In a paper posted online on March 13, 2016, Kannan Soundararajan and Robert Lemke Oliver of Stanford University presented both numerical and theoretical evidence that prime numbers repel other would-be primes that end in the same digit, and have varied predilections for being followed by primes ending in the other possible final digits.[1]

"We've been studying primes for a long time, and no one spotted this before," said Andrew Granville of the University of Montreal. "It's crazy."

The discovery is the exact opposite of what most mathematicians would have predicted, said Ken Ono, a number theorist at Emory University in Atlanta. When he first heard the news, he said, "I was floored. I thought, 'For sure, your program's not working.'"

This conspiracy among prime numbers seems, at first glance, to violate a longstanding assumption in number theory: that prime numbers behave much like random numbers. Most mathematicians would have assumed, Granville and Ono agreed, that a prime should have an equal chance of being followed by a prime ending in 1, 3, 7 or 9 (the four possible endings for all prime numbers except 2 and 5).

"I can't believe anyone in the world would have guessed this," Granville said. Even after having seen Lemke Oliver and Soundararajan's analysis of their phenomenon, he said, "it still seems like a strange thing."

Yet the pair's work doesn't upend the notion that primes behave randomly so much as point to how subtle their particular mix of randomness and order is. "Can we redefine what 'random' means in this context so that once again, [this phenomenon] looks like it might be random?" Soundararajan said. "That's what we think we've done."

PRIME PREFERENCES

Soundararajan was drawn to study consecutive primes after hearing a lecture at Stanford by the mathematician Tadashi Tokieda in which he mentioned a counterintuitive property of coin-tossing: If Alice tosses a coin until she sees a head followed by a tail, and Bob tosses a coin until he sees two heads in a row, then on average, Alice will require four tosses while Bob will require six tosses (try this at home!), even though head-tail and head-head have an equal chance of appearing after two coin tosses.

Soundararajan wondered if similarly strange phenomena appear in other contexts. Since he had studied the primes for decades, he turned to them—and found something even stranger than he had bargained for. Looking at prime numbers written in base 3—in which roughly half the primes end in 1 and half end in 2—he found that among primes smaller than 1,000, a prime ending in 1 is more than twice as likely to be followed by a prime ending in 2 than by another prime ending in 1. Likewise, a prime ending in 2 prefers to be followed a prime ending in 1.

Soundararajan showed his findings to then-postdoctoral researcher Lemke Oliver, who was shocked. He immediately wrote a program that searched much farther out along the number line—through the first 400 billion primes. Lemke Oliver again found that primes seem to avoid being followed by another prime with the same final digit. The primes "really hate to repeat themselves," Lemke Oliver said.

Lemke Oliver and Soundararajan discovered that this sort of bias in the final digits of consecutive primes holds not just in base 3, but also in base 10 and several other bases; they conjecture that it's true in every base. The biases that they found appear to even out, little by little, as you go farther along the number line—but they do so at a snail's pace. "It's the rate at which they even out which is surprising to me," said James Maynard, a number theorist at the University of Oxford. When Soundararajan first told Maynard what the pair had discovered, "I only half believed him," Maynard said. "As soon as I went back to my office, I ran a numerical experiment to check this myself."

Lemke Oliver and Soundararajan's first guess for why this bias occurs was a simple one: Maybe a prime ending in 3, say, is more likely to be followed by a prime ending in 7, 9 or 1 merely because it encounters numbers with those endings before it reaches another number ending in 3. For example,

43 is followed by 47, 49 and 51 before it hits 53, and one of those numbers, 47, is prime.

But the pair of mathematicians soon realized that this potential explanation couldn't account for the magnitude of the biases they found. Nor could it explain why, as the pair found, primes ending in 3 seem to like being followed by primes ending in 9 more than 1 or 7. To explain these and other preferences, Lemke Oliver and Soundararajan had to delve into the deepest model mathematicians have for random behavior in the primes.

RANDOM PRIMES

Prime numbers, of course, are not really random at all—they are completely determined. Yet in many respects, they seem to behave like a list of random numbers, governed by just one overarching rule: The approximate density of primes near any number is inversely proportional to how many digits the number has.

In 1936, the Swedish mathematician Harald Cramér explored this idea using an elementary model for generating random prime-like numbers: At every whole number, flip a weighted coin—weighted by the prime density near that number—to decide whether to include that number in your list of random "primes."[2] Cramér showed that this coin-tossing model does an excellent job of predicting certain features of the real primes, such as how many to expect between two consecutive perfect squares.

Despite its predictive power, Cramér's model is a vast oversimplification. For instance, even numbers have as good a chance of being chosen as odd numbers, whereas real primes are never even, apart from the number 2. Over the years, mathematicians have developed refinements of Cramér's model that, for instance, bar even numbers and numbers divisible by 3, 5 and other small primes.

These simple coin-tossing models tend to be very useful rules of thumb about how prime numbers behave. They accurately predict, among other things, that prime numbers shouldn't care what their final digit is—and indeed, primes ending in 1, 3, 7 and 9 occur with roughly equal frequency.

Yet similar logic seems to suggest that primes shouldn't care what digit the prime after them ends in. It was probably mathematicians' overreliance on the simple coin-tossing heuristics that made them miss the biases in consecutive primes for so long, Granville said. "It's easy to take too much for granted—to assume that your first guess is true."

The primes' preferences about the final digits of the primes that follow them can be explained, Soundararajan and Lemke Oliver found, using a much more refined model of randomness in primes, something called the prime k-tuples conjecture. Originally stated by mathematicians G. H. Hardy and J. E. Littlewood in 1923, the conjecture provides precise estimates of how often every possible constellation of primes with a given spacing pattern will appear.[3] A wealth of numerical evidence supports the conjecture, but so far a proof has eluded mathematicians.

The prime k-tuples conjecture subsumes many of the most central open problems in prime numbers, such as the twin primes conjecture, which posits that there are infinitely many pairs of primes—such as 17 and 19—that are only two apart. Most mathematicians believe the twin primes conjecture not so much because they keep finding more twin primes, Maynard said, but because the number of twin primes they've found fits so neatly with what the prime k-tuples conjecture predicts.

In a similar way, Soundararajan and Lemke Oliver have found that the biases they uncovered in consecutive primes come very close to what the prime k-tuples conjecture predicts. In other words, the most sophisticated conjecture mathematicians have about randomness in primes forces the primes to display strong biases. "I have to rethink how I teach my class in analytic number theory now," Ono said.

At this early stage, mathematicians say, it's hard to know whether these biases are isolated peculiarities, or whether they have deep connections to other mathematical structures in the primes or elsewhere. Ono predicts, however, that mathematicians will immediately start looking for similar biases in related contexts, such as prime polynomials—fundamental objects in number theory that can't be factored into simpler polynomials.

And the finding will make mathematicians look at the primes themselves with fresh eyes, Granville said. "You could wonder, what else have we missed about the primes?"

II   IS MATH THE UNIVERSAL LANGUAGE OF NATURE?

MATHEMATICIANS CHASE MOONSHINE'S SHADOW

Erica Klarreich

I n 1978, the mathematician John McKay noticed what seemed like an odd coincidence. He had been studying the different ways of representing the structure of a mysterious entity called the monster group, a gargantuan algebraic object that, mathematicians believed, captured a new kind of symmetry. Mathematicians weren't sure that the monster group actually existed, but they knew that if it did exist, it acted in special ways in particular dimensions, the first two of which were 1 and 196,883.

McKay, of Concordia University in Montreal, happened to pick up a mathematics paper in a completely different field, involving something called the j-function, one of the most fundamental objects in number theory. Strangely enough, this function's first important coefficient is 196,884, which McKay instantly recognized as the sum of the monster's first two special dimensions.

Most mathematicians dismissed the finding as a fluke, since there was no reason to expect the monster and the j-function to be even remotely related. However, the connection caught the attention of John Thompson, a Fields medalist now at the University of Florida in Gainesville, who made an additional discovery. The j-function's second coefficient, 21,493,760, is the sum of the first three special dimensions of the monster: $1 + 196,883 + 21,296,876$. It seemed as if the j-function was somehow controlling the structure of the elusive monster group.

Soon, two other mathematicians had demonstrated so many of these numerical relationships that it no longer seemed possible that they were mere coincidences. In a 1979 paper called "Monstrous Moonshine," the pair—John Conway, now an emeritus professor at Princeton University, and Simon Norton—conjectured that these relationships must result from some deep connection between the monster group and the j-function.[1] "They called it moonshine because it appeared so far-fetched," said Don Zagier, a director of the Max Planck Institute for Mathematics in Bonn, Germany. "They were such wild ideas that it seemed like wishful thinking to imagine anyone could ever prove them."

It took several more years before mathematicians succeeded in even constructing the monster group, but they had a good excuse: The monster has more than 10^{53} elements, which is more than the number of atoms in a thousand Earths.[2] In 1992, a decade after Robert Griess of the University of Michigan constructed the monster, Richard Borcherds tamed the wild ideas of monstrous moonshine, eventually earning a Fields Medal for this work.[3] Borcherds, of the University of California, Berkeley, proved that there was a bridge between the two distant realms of mathematics in which the monster and the j-function live: namely, string theory, the counterintuitive idea that the universe has tiny hidden dimensions, too small to measure, in which strings vibrate to produce the physical effects we experience at the macroscopic scale.

Borcherds' discovery touched off a revolution in pure mathematics, leading to a new field known as generalized Kac–Moody algebras. But from a string theory point of view, it was something of a backwater. The 24-dimensional string theory model that linked the j-function and the monster was far removed from the models string theorists were most excited about. "It seemed like just an esoteric corner of the theory, without much physical interest, although the math results were startling," said Shamit Kachru, a string theorist at Stanford University.

But now moonshine is undergoing a renaissance, one that may eventually have deep implications for string theory. Over the past eight years, starting with a discovery analogous to McKay's, mathematicians and physicists have come to realize that monstrous moonshine is just the start of the story.

In March 2015, researchers posted a paper on arxiv.org presenting a numerical proof of the so-called Umbral Moonshine Conjecture, formulated in 2012, which proposes that in addition to monstrous moonshine, there are 23 other moonshines: mysterious correspondences between the dimensions of a symmetry group on the one hand, and the coefficients of a special function on the other.[4] The functions in these new moonshines have their origins in a prescient letter by one of mathematics' great geniuses, written more than half a century before moonshine was even a glimmer in the minds of mathematicians.

The 23 new moonshines appear to be intertwined with some of the most central structures in string theory, four-dimensional objects known as K3 surfaces. The connection with umbral moonshine hints at hidden symmetries in these surfaces, said Miranda Cheng of the University of Amsterdam and France's National Center for Scientific Research, who originated the Umbral Moonshine Conjecture together with John Duncan, then at Case Western Reserve University in Cleveland, Ohio, and Jeffrey Harvey, of the University of Chicago. "This is important, and we need to understand it," she said.

The new proof strongly suggests that in each of the 23 cases, there must be a string theory model that holds the key to understanding these otherwise baffling numerical correspondences. But the proof doesn't go so far as to actually construct the relevant string theory models, leaving physicists with a tantalizing problem. "At the end of the day when we understand what moonshine is, it will be in terms of physics," Duncan said.

MONSTROUS MOONSHINE

The symmetries of any given shape have a natural sort of arithmetic to them. For example, rotating a square 90 degrees and then flipping it horizontally is the same as flipping it across a diagonal—in other words, "90-degree rotation + horizontal flip = diagonal flip." During the 19th century, mathematicians realized that they could distill this type of arithmetic into an algebraic entity called a group. The same abstract group can represent the symmetries of many different shapes, giving mathematicians a tidy way to understand the commonalities in different shapes.

Over much of the 20th century, mathematicians worked to classify all possible groups, and they gradually discovered something strange: While most simple finite groups fell into natural categories, there were 26 oddballs that defied categorization. Of these, the biggest, and the last to be discovered, was the monster.

Before McKay's serendipitous discovery nearly four decades ago, there was no reason to think the monster group had anything to do with the *j*-function, the second protagonist of the monstrous-moonshine story. The *j*-function belongs to a special class of functions whose graphs have repeating patterns similar to M. C. Escher's tessellation of a disk with angels and devils, which shrink ever smaller as they approach the outer boundary. These "modular" functions are the heroes of number theory, playing a crucial role, for instance, in Andrew Wiles' 1994 proof of Fermat's Last Theorem.[5] "Any time you hear about a striking result in number theory, there's a high chance that it's really a statement about modular forms," Kachru said.

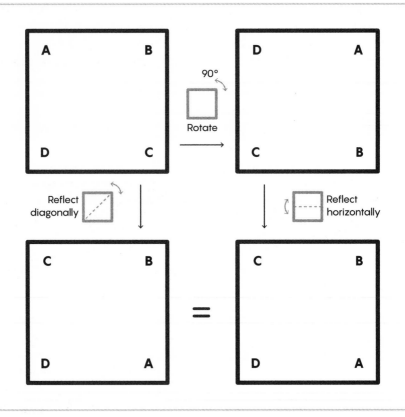

FIGURE 2.1

Rotating a square 90 degrees and then reflecting it horizontally has the same effect as reflecting it across a diagonal, so in the language of square symmetry arithmetic, 90-degree rotation+horizontal reflection=diagonal reflection.

As with a sound wave, the *j*-function's repeating pattern can be broken down into a collection of pure tones, so to speak, with coefficients indicating how "loud" each tone is. It is in these coefficients that McKay found the link to the monster group.

In the early 1990s, building on work by Igor Frenkel of Yale University, James Lepowsky of Rutgers University and Arne Meurman of Lund University in Sweden, Borcherds made sense of McKay's discovery by showing that there is a particular string theory model in which the *j*-function and the monster group both play roles. The coefficients of the *j*-function count the ways strings can oscillate at each energy level. And the monster group captures the model's symmetry at those energy levels.

The finding gave mathematicians a way to study the mind-bogglingly large monster group using the j-function, whose coefficients are easy to calculate. "Math is all about building bridges where on one side you see more clearly than on the other," Duncan said. "But this bridge was so unexpectedly powerful that before you see the proof it's kind of crazy."

NEW MOONSHINE

While mathematicians explored the ramifications of monstrous moonshine, string theorists focused on a seemingly different problem: figuring out the geometry for the tiny dimensions in which strings are hypothesized to live. Different geometries allow strings to vibrate in different ways, just as tightening the tension on a drum changes its pitch. For decades, physicists have struggled to find a geometry that produces the physical effects we see in the real world.

An important ingredient in some of the most promising candidates for such a geometry is a collection of four-dimensional shapes known as K3 surfaces. In contrast with Borcherds' string theory model, Kachru said, K3 surfaces fill the string theory textbooks.

Not enough is known about the geometry of K3 surfaces to count how many ways strings can oscillate at each energy level, but physicists can write down a more limited function counting certain physical states that appear in all K3 surfaces. In 2010, three string theorists—Tohru Eguchi of Kyoto University in Japan, Hirosi Ooguri of the California Institute of Technology in Pasadena and Yuji Tachikawa of the University of Tokyo in Japan—noticed that if they wrote this function in a particular way, out popped coefficients that were the same as some special dimensions of another oddball group, called the Mathieu 24 (M24) group, which has nearly 250 million elements.[6] The three physicists had discovered a new moonshine.

This time, physicists and mathematicians were all over the discovery. "I was at several conferences, and all the talk was about this new Mathieu moonshine," Zagier said.

Zagier attended one such conference in Zurich in July 2011, and there, Duncan wrote in an email, Zagier showed him "a piece of paper with lots of numbers on it"—the coefficients of some functions Zagier was studying called "mock modular" forms, which are related to modular functions. "Don [Zagier] pointed to a particular line of numbers and asked me—in jest, I think—if there is any finite group related to them," Duncan wrote.

Duncan wasn't sure, but he recognized the numbers on another line: They belonged to the special dimensions of a group called M12. Duncan buttonholed Miranda Cheng, and the two pored over the rest of Zagier's piece of paper. The pair, together with Jeffrey Harvey, gradually realized that there was much more to the new moonshine than just the M24 example. The clue to the full moonshine picture, they found, lay in the nearly century-old writings of one of mathematics' legendary figures.

MOONSHINE'S SHADOWS

In 1913, the English mathematician G. H. Hardy received a letter from an accounting clerk in Madras, India, describing some mathematical formulas he had discovered. Many of them were old hat, and some were flat-out wrong, but on the final page were three formulas that blew Hardy's mind. "They must be true," wrote Hardy, who promptly invited the clerk, Srinivasa Ramanujan, to England, "because, if they were not true, no one would have the imagination to invent them."

Ramanujan became famous for seemingly pulling mathematical relationships out of thin air, and he credited many of his discoveries to the goddess Namagiri, who appeared to him in visions, he said. His mathematical career was tragically brief, and in 1920, as he lay dying in India at age 32, he wrote Hardy another letter saying that he had discovered what he called "mock theta" functions, which entered into mathematics "beautifully." Ramanujan listed 17 examples of these functions, but didn't explain what they had in common. The question remained open for more than eight decades, until Sander Zwegers, then a graduate student of Zagier's and now a professor at the University of Cologne in Germany, figured out in 2002 that they are all examples of what came to be known as mock modular forms.[7]

After the Zurich moonshine conference, Cheng, Duncan and Harvey gradually figured out that M24 moonshine is one of 23 different moonshines, each making a connection between the special dimensions of a group and the coefficients of a mock modular form—just as monstrous moonshine made a connection between the monster group and the *j*-function. For each of these moonshines, the researchers conjectured, there is a string theory like the one in monstrous moonshine, in which the mock modular form counts the string states and the group captures the model's symmetry. A mock modular form always has an associated modular function called its "shadow," so they named their hypothesis the Umbral Moonshine Conjecture—*umbra* is Latin for "shadow." Many of the mock modular forms that appear in the conjecture are among the 17 special examples Ramanujan listed in his prophetic letter.

Curiously enough, Borcherds' earlier proof of monstrous moonshine also builds on work by Ramanujan: The algebraic objects at the core of the proof were discovered by Frenkel, Lepowsky and Meurman as they analyzed the three formulas that had so startled Hardy in Ramanujan's first letter. "It's amazing that these two letters form the cornerstone of what we know about moonshine," said Ken Ono, of Emory University in Atlanta, Georgia. "Without either letter, we couldn't write this story."

FINDING THE BEAST

In the 2015 paper posted on arxiv.org, Duncan, Ono and Ono's then–graduate student Michael Griffin came up with a numerical proof of the Umbral Moonshine Conjecture (one case of which—the M24 case—had already been proven by Terry Gannon, of the University of Alberta in Edmonton, Canada).[8] The new analysis provides only hints of where physicists should look for the string theories that will unite the groups and the mock modular forms. Nevertheless, the proof confirms that the conjecture is on the right track, Harvey said. "We had all this structure, and it was so intricate and compelling that it was hard not to think there was some truth to it," he said. "Having a mathematical proof makes it a solid piece of work that people can think seriously about."

The string theory underlying umbral moonshine is likely to be "not just any physical theory, but a particularly important one," Cheng said. "It suggests that there's a special symmetry acting on the physical theory of K3 surfaces." Researchers studying K3 surfaces can't see this symmetry yet, she said, suggesting that "there is probably a better way of looking at that theory that we haven't found yet."

Physicists are also excited about a highly conjectural connection between moonshine and quantum gravity, the as-yet-undiscovered theory that will unite general relativity and quantum mechanics. In 2007, the physicist Edward Witten, of the Institute for Advanced Study, speculated that the string theory in monstrous moonshine should offer a way to construct a model of three-dimensional quantum gravity, in which 194 natural categories of elements in the monster group correspond to 194 classes of black holes.[9] Umbral moonshine may lead physicists to similar conjectures, giving hints of where to look for a quantum gravity theory. "That is a big hope for the field," Duncan said.

The numerical proof of the Umbral Moonshine Conjecture is "like looking for an animal on Mars and seeing its footprint, so we know it's there," Zagier said. Now, researchers have to find the animal—the string theory that would illuminate all these deep connections. "We really want to get our hands on it," Zagier said.

IN MYSTERIOUS PATTERN, MATH AND NATURE CONVERGE

Natalie Wolchover

I n 1999, while sitting at a bus stop in Cuernavaca, Mexico, a Czech physicist named Petr Šeba noticed young men handing slips of paper to the bus drivers in exchange for cash. It wasn't organized crime, he learned, but another shadow trade: Each driver paid a "spy" to record when the bus ahead of his had departed the stop. If it had left recently, he would slow down, letting passengers accumulate at the next stop. If it had departed long ago, he sped up to keep other buses from passing him. This system maximized profits for the drivers. And it gave Šeba an idea.

"We felt here some kind of similarity with quantum chaotic systems," explained Šeba's co-author, Milan Krbálek, in an email.

After several failed attempts to talk to the spies himself, Šeba asked his student to explain to them that he wasn't a tax collector, or a criminal—he was simply a "crazy" scientist willing to trade tequila for their data. The men handed over their used papers. When the researchers plotted thousands of bus departure times on a computer, their suspicions were confirmed: The interaction between drivers caused the spacing between departures to exhibit a distinctive pattern previously observed in quantum physics experiments.

"I was thinking that something like this could come out, but I was really surprised that it comes exactly," Šeba said.

Subatomic particles have little to do with decentralized bus systems. But in the years since the odd coupling was discovered, the same pattern has turned up in other unrelated settings. Scientists now believe the widespread phenomenon, known as "universality," stems from an underlying connection to mathematics, and it is helping them to model complex systems from the internet to Earth's climate.

The pattern was first discovered in nature in the 1950s in the energy spectrum of the uranium nucleus, a behemoth with hundreds of moving parts

RANDOM

UNIVERSALITY

PERIODIC

FIGURE 2.2

The light gray pattern exhibits a precise balance of randomness and regularity known as "universality," which has been observed in the spectra of many complex, correlated systems. In this spectrum, a mathematical formula called the "correlation function" gives the exact probability of finding two lines spaced a given distance apart.

that quivers and stretches in infinitely many ways, producing an endless sequence of energy levels. In 1972, the number theorist Hugh Montgomery observed it in the zeros of the Riemann zeta function, a mathematical object closely related to the distribution of prime numbers.[1] In 2000, Krbálek and Šeba reported it in the Cuernavaca bus system.[2] And in more recent years it has shown up in spectral measurements of composite materials, such as sea ice and human bones, and in signal dynamics of the Erdős–Rényi model, a simplified version of the internet named for Paul Erdős and Alfréd Rényi.[3]

Each of these systems has a spectrum—a sequence like a bar code representing data such as energy levels, zeta zeros, bus departure times or signal speeds. In all the spectra, the same distinctive pattern appears: The data seem haphazardly distributed, and yet neighboring lines repel one another, lending a degree of regularity to their spacing. This fine balance between chaos and order, which is defined by a precise formula, also appears in a purely mathematical setting: It defines the spacing between the eigenvalues, or solutions, of a vast matrix filled with random numbers.

"Why so many physical systems behave like random matrices is still a mystery," said Horng-Tzer Yau, a mathematician at Harvard University. But in recent years, he said, "we have made a very important step in our understanding."

By investigating the "universality" phenomenon in random matrices, researchers have developed a better sense of why it arises elsewhere—and how it can be used. In a flurry of papers, Yau and other mathematicians have characterized many new types of random matrices, which can conform to a variety of numerical distributions and symmetry rules. For example, the numbers filling a matrix's rows and columns might be chosen from a bell curve of possible values, or they might simply be 1s and –1s. The top right and bottom left halves of the matrix might be mirror images of one another, or not. Time and again, regardless of their specific characteristics, the random matrices are found to exhibit that same chaotic yet regular pattern in the distribution of their eigenvalues. That's why mathematicians call the phenomenon "universality."

"It seems to be a law of nature," said Van Vu, a mathematician at Yale University who, with Terence Tao of UCLA, has proven universality for a broad class of random matrices.

Universality is thought to arise when a system is very complex, consisting of many parts that strongly interact with each other to generate a spectrum. The pattern emerges in the spectrum of a random matrix, for example, because the matrix elements all enter into the calculation of that spectrum. But random matrices are merely "toy systems" that are of interest because they can be rigorously studied, while also being rich enough to model real-world systems, Vu said. Universality is much more widespread. Wigner's hypothesis (named after Eugene Wigner, the physicist who discovered universality in atomic spectra) asserts that all complex, correlated systems exhibit universality, from a crystal lattice to the internet.

The more complex a system is, the more robust its universality should be, said László Erdős of the University of Munich, one of Yau's collaborators. "This is because we believe that universality is the typical behavior."

In many simple systems, individual components can assert too great an influence on the outcome of the system, changing the spectral pattern. With larger systems, no single component dominates. "It's like if you have a room with a lot of people and they decide to do something, the personality of one person isn't that important," Vu said.

Whenever a system exhibits universality, the behavior acts as a signature certifying that the system is complex and correlated enough to be treated like a random matrix. "This means you can use a random matrix to model it," Vu said. "You can compute other parameters of the matrix model and use them to predict that the system may behave like the parameters you computed."

This technique is enabling scientists to understand the structure and evolution of the internet. Certain properties of this vast computer network, such as the typical size of a cluster of computers, can be closely estimated by measurable properties of the corresponding random matrix. "People are very interested in clusters and their locations, partially motivated by practical purposes such as advertising," Vu said.

A similar technique may lead to improvements in climate change models. Scientists have found that the presence of universality in features similar to the energy spectrum of a material indicates that its components are highly connected, and that it will therefore conduct fluids, electricity or heat. Conversely, the absence of universality may show that a material is sparse and acts as an insulator. In work presented in January 2013 at the Joint Mathematics Meetings in San Diego, California, Ken Golden, a mathematician at the University of Utah, and his student, Ben Murphy, used this distinction to predict heat transfer and fluid flow in sea ice, both at the microscopic level and through patchworks of Arctic melt ponds spanning thousands of kilometers.[4]

The spectral measure of a mosaic of melt ponds, taken from a helicopter, or a similar measurement taken of a sample of sea ice in an ice core, instantly exposes the state of either system. "Fluid flow through sea ice governs or mediates very important processes that you need to understand in order to understand the climate system," Golden said. "The transitions in the eigenvalue statistics present a brand new, mathematically rigorous approach to incorporating sea ice into climate models."

The same trick may also eventually provide an easy test for osteoporosis. Golden, Murphy and their colleagues have found that the spectrum of a dense, healthy bone exhibits universality, while that of a porous, osteoporotic bone does not.

"We're dealing with systems where the 'particles' can be on the millimeter or even on the kilometer scale," Murphy said, referring to the systems' component parts. "It's amazing that the same underlying mathematics describes both."

The reason a real-world system would exhibit the same spectral behavior as a random matrix may be easiest to understand in the case of the nucleus of a heavy atom. All quantum systems, including atoms, are governed by the rules of mathematics, and specifically by those of matrices. "That's what quantum mechanics is all about," said Freeman Dyson, a retired mathematical physicist who helped develop random matrix theory in the 1960s and 1970s while at the Institute for Advanced Study. "Every quantum system is governed by a matrix representing the total energy of the system, and the eigenvalues of the matrix are the energy levels of the quantum system."

The matrices behind simple atoms, such as hydrogen or helium, can be worked out exactly, yielding eigenvalues that correspond with stunning precision to the measured energy levels of the atoms. But the matrices corresponding to more complex quantum systems, such as a uranium nucleus, quickly grow too thorny to grasp. According to Dyson, this is why such nuclei can be compared to random matrices. Many of the interactions inside uranium—the elements of its unknown matrix—are so complex that they become washed out, like a mélange of sounds blending into noise. Consequently, the unknown matrix that governs the nucleus behaves like a matrix filled with random numbers, and so its spectrum exhibits universality.

Scientists have yet to develop an intuitive understanding of why this particular random-yet-regular pattern, and not some other pattern, emerges for complex systems. "We only know it from calculations," Vu said. Another mystery is what it has to do with the Riemann zeta function, whose spectrum of zeros exhibits universality. The zeros of the zeta function are closely tied to the distribution of the prime numbers—the irreducible integers out of which all others are constructed. Mathematicians have long wondered at the haphazard way in which the primes are sprinkled along the number line from one to infinity, and universality offers a clue. Some think there may be a matrix underlying the Riemann zeta function that is complex and correlated enough to exhibit universality. Discovering such a matrix would have "big implications" for finally understanding the distribution of the primes, said the mathematician Paul Bourgade.

Or perhaps the explanation lies deeper still. "It may happen that it is not a matrix that lies at the core of both Wigner's universality and the zeta function, but some other, yet undiscovered, mathematical structure," Erdős said. "Wigner matrices and zeta functions may then just be different representations of this structure."

Many mathematicians are searching for the answer, with no guarantee that there is one. "Nobody imagined that the buses in Cuernavaca would turn out to be an example of this. Nobody imagined that the zeroes of the zeta function would be another example," Dyson said. "The beauty of science is it's completely unpredictable, and so everything useful comes out of surprises."

AT THE FAR ENDS OF A NEW UNIVERSAL LAW

Natalie Wolchover

I magine an archipelago where each island hosts a single tortoise species and all the islands are connected—say by rafts of flotsam. As the tortoises interact by dipping into one another's food supplies, their populations fluctuate.

In 1972, the biologist Robert May devised a simple mathematical model that worked much like the archipelago. He wanted to figure out whether a complex ecosystem can ever be stable or whether interactions between species inevitably lead some to wipe out others. By indexing chance interactions between species as random numbers in a matrix, he calculated the critical "interaction strength"—a measure of the number of flotsam rafts, for example—needed to destabilize the ecosystem.[1] Below this critical point, all species maintained steady populations. Above it, the populations shot toward zero or infinity.

Little did May know, the tipping point he discovered was one of the first glimpses of a curiously pervasive statistical law.

The law appeared in full form two decades later, when the mathematicians Craig Tracy and Harold Widom proved that the critical point in the kind of model May used was the peak of a statistical distribution. Then, in 1999, Jinho Baik, Percy Deift and Kurt Johansson discovered that the same statistical distribution also describes variations in sequences of shuffled integers—a completely unrelated mathematical abstraction. Soon the distribution appeared in models of the wriggling perimeter of a bacterial colony and other kinds of random growth. Before long, it was showing up all over physics and mathematics.

"The big question was why," said Satya Majumdar, a statistical physicist at the University of Paris-Sud. "Why does it pop up everywhere?"

Systems of many interacting components—be they species, integers or subatomic particles—kept producing the same statistical curve, which had become known as the Tracy–Widom distribution. This puzzling curve seemed to be the complex cousin of the familiar bell curve, or Gaussian distribution, which represents the natural variation of independent random variables like the heights of students in a classroom or their test scores. Like the Gaussian, the Tracy–Widom distribution exhibits "universality," a mysterious phenomenon in which diverse microscopic effects give rise to the same collective behavior. "The surprise is it's as universal as it is," said Tracy, a professor at the University of California, Davis.

When uncovered, universal laws like the Tracy–Widom distribution enable researchers to accurately model complex systems whose inner workings they know little about, like financial markets, exotic phases of matter or the internet.

"It's not obvious that you could have a deep understanding of a very complicated system using a simple model with just a few ingredients," said Grégory Schehr, a statistical physicist who works with Majumdar at Paris-Sud. "Universality is the reason why theoretical physics is so successful."

Universality is "an intriguing mystery," said Terence Tao of UCLA. Why do certain laws seem to emerge from complex systems, he asked, "almost regardless of the underlying mechanisms driving those systems at the microscopic level?"

Now, through the efforts of researchers like Majumdar and Schehr, a surprising explanation for the ubiquitous Tracy–Widom distribution is beginning to emerge.

LOPSIDED CURVE

The Tracy–Widom distribution is an asymmetrical statistical bump, steeper on the left side than the right. Suitably scaled, its summit sits at a telltale value: $\sqrt{2N}$, the square root of twice the number of variables in the systems that give rise to it and the exact transition point between stability and instability that May calculated for his model ecosystem.

The transition point corresponded to a property of his matrix model called the "largest eigenvalue": the greatest in a series of numbers calculated from the matrix's rows and columns. Researchers had already discovered that the N eigenvalues of a "random matrix"—one filled with random numbers—tend to space apart along the real number line according to a distinct pattern, with the largest eigenvalue typically located at or near $\sqrt{2N}$. Tracy and Widom determined how the largest eigenvalues of random matrices fluctuate around this average value, piling up into the lopsided statistical distribution that bears their names.

GAUSSIAN DISTRIBUTION
Uncorrelated variables

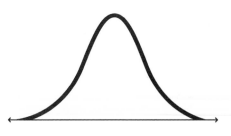

TRACY — WIDOM DISTRIBUTION
Correlated variables

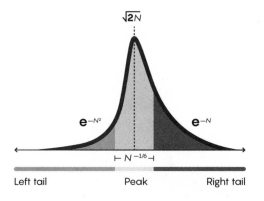

FIGURE 2.3
Whereas "uncorrelated" random variables such as test scores splay out into the bell-shaped Gaussian distribution, interacting species, financial stocks and other "correlated" variables give rise to a more complicated statistical curve. Steeper on the left than the right, the curve has a shape that depends on N, the number of variables.

When the Tracy–Widom distribution turned up in the integer sequences problem and other contexts that had nothing to do with random matrix theory, researchers began searching for the hidden thread tying all its manifestations together, just as mathematicians in the 18th and 19th centuries sought a theorem that would explain the ubiquity of the bell-shaped Gaussian distribution.

The central limit theorem, which was finally made rigorous about a century ago, certifies that test scores and other "uncorrelated" variables— meaning any of them can change without affecting the rest—will form a bell curve. By contrast, the Tracy–Widom curve appears to arise from variables that are strongly correlated, such as interacting species, stock prices and matrix eigenvalues. The feedback loop of mutual effects between correlated variables makes their collective behavior more complicated than that of uncorrelated variables like test scores. While researchers have rigorously proved certain classes of random matrices in which the Tracy–Widom distribution universally holds, they have a looser handle on its manifestations in counting problems, random-walk problems, growth models and beyond.[2]

"No one really knows what you need in order to get Tracy–Widom," said Herbert Spohn, a mathematical physicist at the Technical University of Munich in Germany. "The best we can do," he said, is to gradually uncover the range of its universality by tweaking systems that exhibit the distribution and seeing whether the variants give rise to it too.

So far, researchers have characterized three forms of the Tracy–Widom distribution: rescaled versions of one another that describe strongly correlated systems with different types of inherent randomness. But there could be many more than three, perhaps even an infinite number, of Tracy–Widom universality classes. "The big goal is to find the scope of universality of the Tracy–Widom distribution," said Baik, a professor of mathematics at the University of Michigan. "How many distributions are there? Which cases give rise to which ones?"

As other researchers identified further examples of the Tracy–Widom peak, Majumdar, Schehr and their collaborators began hunting for clues in the curve's left and right tails.

GOING THROUGH A PHASE

Majumdar became interested in the problem in 2006 during a workshop at the University of Cambridge in England. He met a pair of physicists who were using random matrices to model string theory's abstract space of all possible universes. The string theorists reasoned that stable points in this "landscape" corresponded to the subset of random matrices whose largest eigenvalues were negative—far to the left of the average value of $\sqrt{2N}$ at the peak of the Tracy–Widom curve.[3] They wondered just how rare these stable points—the seeds of viable universes—might be.

To answer the question, Majumdar and David Dean, now of the University of Bordeaux in France, realized that they needed to derive an equation describing the tail to the extreme left of the Tracy–Widom peak, a region of the statistical distribution that had never been studied. Within a year, their derivation of the left "large deviation function" appeared in *Physical Review Letters*.[4] Using different techniques, Majumdar and Massimo Vergassola, now of the University of California, San Diego, calculated the right large deviation function three years later. On the right, Majumdar and Dean were surprised to find that the distribution dropped off at a rate related to the number of eigenvalues, N; on the left, it tapered off more quickly, as a function of N^2.

In 2011, the form of the left and right tails gave Majumdar, Schehr and Peter Forrester of the University of Melbourne in Australia a flash of insight: They realized the universality of the Tracy–Widom distribution could be related to the universality of phase transitions—events such as water freezing into ice, graphite becoming diamond and ordinary metals transforming into strange superconductors.

Because phase transitions are so widespread—all substances change phases when fed or starved of sufficient energy—and take only a handful of mathematical forms, they are for statistical physicists "almost like a religion," Majumdar said.

In the miniscule margins of the Tracy–Widom distribution, Majumdar, Schehr and Forrester recognized familiar mathematical forms: distinct curves describing two different rates of change in the properties of a system, sloping downward from either side of a transitional peak. These were the trappings of a phase transition.

In the thermodynamic equations describing water, the curve that represents the water's energy as a function of temperature has a kink at 100 degrees Celsius, the point at which the liquid becomes steam. The water's energy slowly increases up to this point, suddenly jumps to a new level and then slowly increases again along a different curve, in the form of steam. Crucially, where the energy curve has a kink, the "first derivative" of the curve—another curve that shows how quickly the energy changes at each point—has a peak.

Similarly, the physicists realized, the energy curves of certain strongly correlated systems have a kink at $\sqrt{2N}$. The associated peak for these systems is the Tracy–Widom distribution, which appears in the third derivative of the energy curve—that is, the rate of change of the rate of change of the energy's rate of change. This makes the Tracy–Widom distribution a "third-order" phase transition.

"The fact that it pops up everywhere is related to the universal character of phase transitions," Schehr said. "This phase transition is universal in the sense that it does not depend too much on the microscopic details of your system."

According to the form of the tails, the phase transition separated phases of systems whose energy scaled with N^2 on the left and N on the right. But Majumdar and Schehr wondered what characterized this Tracy–Widom universality class; why did third-order phase transitions always seem to occur in systems of correlated variables?

The answer lay buried in a pair of esoteric papers from 1980. A third-order phase transition had shown up before, identified that year in a simplified version of the theory governing atomic nuclei. The theoretical physicists David Gross, Edward Witten and (independently) Spenta Wadia discovered a third-order phase transition separating a "weak coupling" phase, in which matter takes the form of nuclear particles, and a higher-temperature "strong coupling" phase, in which matter melds into plasma.[5] After the Big Bang, the universe probably transitioned from a strong- to a weak-coupling phase as it cooled.

After examining the literature, Schehr said, he and Majumdar "realized there was a deep connection between our probability problem and this third-order phase transition that people had found in a completely different context."

WEAK TO STRONG

Majumdar and Schehr have since accrued substantial evidence that the Tracy–Widom distribution and its large deviation tails represent a universal phase transition between weak- and strong-coupling phases.[6] In May's ecosystem model, for example, the critical point at $\sqrt{2N}$ separates a stable phase of weakly coupled species, whose populations can fluctuate individually without affecting the rest, from an unstable phase of strongly coupled species, in which fluctuations cascade through the ecosystem and throw it off balance. In general, Majumdar and Schehr believe, systems in the Tracy–Widom universality class exhibit one phase in which all components act in concert and another phase in which the components act alone.

The asymmetry of the statistical curve reflects the nature of the two phases. Because of mutual interactions between the components, the energy of the system in the strong-coupling phase on the left is proportional to N^2. Meanwhile, in the weak-coupling phase on the right, the energy depends only on the number of individual components, N.

"Whenever you have a strongly coupled phase and a weakly coupled phase, Tracy–Widom is the connecting crossover function between the two phases," Majumdar said.

Majumdar and Schehr's work is "a very nice contribution," said Pierre Le Doussal, a physicist at École Normale Supérieure in France who helped prove the presence of the Tracy–Widom distribution in a stochastic growth model called the KPZ equation.[7] Rather than focusing on the peak of the Tracy–Widom distribution, "the phase transition is probably the deeper level" of explanation, Le Doussal said. "It should basically make us think more about trying to classify these third-order transitions."

Leo Kadanoff, the statistical physicist who introduced the term "universality" and helped classify universal phase transitions in the 1960s, said it had long been clear to him that universality in random matrix theory must somehow be connected to the universality of phase transitions. But while the physical equations describing phase transitions seem to match reality, many of the computational methods used to derive them have never been made mathematically rigorous.

"Physicists will, in a pinch, settle for a comparison with nature," Kadanoff said, "Mathematicians want proofs—proof that phase-transition theory is correct; more detailed proofs that random matrices fall into the universality class of third-order phase transitions; proof that such a class exists."

For the physicists involved, a preponderance of evidence will suffice. The task now is to identify and characterize strong- and weak-coupling phases in more of the systems that exhibit the Tracy–Widom distribution, such as growth models, and to predict and study new examples of Tracy–Widom universality throughout nature.

The telltale sign will be the tails of the statistical curves. At a gathering of experts in Kyoto, Japan, in August 2014, Le Doussal encountered Kazumasa Takeuchi, a University of Tokyo physicist who reported in 2010 that the interface between two phases of a liquid crystal material varies according to the Tracy–Widom distribution.[8] In 2010, Takeuchi had not collected enough data to plot extreme statistical outliers, such as prominent spikes along the interface. But when Le Doussal entreated Takeuchi to plot the data again, the scientists saw the first glimpse of the left and right tails. Le Doussal immediately emailed Majumdar with the news.

"Everybody looks only at the Tracy–Widom peak," Majumdar said. "They don't look at the tails because they are very, very tiny things."

A BIRD'S-EYE VIEW OF NATURE'S HIDDEN ORDER

Natalie Wolchover

Nine years ago, Joe Corbo stared into the eye of a chicken and saw something astonishing. The color-sensitive cone cells that carpeted the retina (detached from the fowl and mounted under a microscope) appeared as polka dots of five different colors and sizes. But Corbo observed that, unlike the randomly dispersed cones in human eyes, or the neat rows of cones in the eyes of many fish, the chicken's cones had a haphazard and yet remarkably uniform distribution. The dots' locations followed no discernible rule, and yet dots never appeared too close together or too far apart. Each of the five interspersed sets of cones, and all of them together, exhibited this same arresting mix of randomness and regularity. Corbo, who runs a biology lab at Washington University in St. Louis, was hooked.

"It's extremely beautiful just to look at these patterns," he said. "We were kind of captured by the beauty, and had, purely out of curiosity, the desire to understand the patterns better." He and his collaborators also hoped to figure out the patterns' function and how they were generated. He didn't know then that these same questions were being asked in numerous other contexts, or that he had found the first biological manifestation of a type of hidden order that has also turned up all over mathematics and physics.

Corbo did know that whatever bird retinas are doing is probably the thing to do. Avian vision works spectacularly well (enabling eagles, for instance, to spot mice from a mile high), and his lab studies the evolutionary adaptations that make this so. Many of these attributes are believed to have been passed down to birds from a lizardlike creature that, 300 million years ago, gave rise to both dinosaurs and proto-mammals. While birds' ancestors, the dinos, ruled the planetary roost, our mammalian kin scurried around in the dark, fearfully nocturnal and gradually losing color discrimination. Mammals' cone types dropped to two—a nadir from which we are still clambering back. About 30 million years ago, one of our primate ancestors' cones split into two—red- and green-detecting—which, together with the existing blue-detecting cone, give us trichromatic vision. But our cones, particularly the newer red and green ones, have a clumpy, scattershot distribution and sample light unevenly.

Bird eyes have had eons longer to optimize. Along with their higher cone count, they achieve a far more regular spacing of the cells. But why, Corbo and colleagues wondered, had evolution not opted for the perfect regularity of a grid or "lattice" distribution of cones? The strange, uncategorizable pattern they observed in the retinas was, in all likelihood, optimizing some unknown set of constraints. What these were, what the pattern was, and how the avian visual system achieved it remained unclear. The biologists did their best to quantify the regularity in the retinas, but this was unfamiliar terrain, and they needed help. In 2012, Corbo contacted Salvatore Torquato, a professor of theoretical chemistry at Princeton University and a renowned expert in a discipline known as "packing." Packing problems ask about the densest way to pack objects (such as cone cells of five different sizes) in a given number of dimensions (in the case of a retina, two). "I wanted to get at this question of whether such a system was optimally packed," Corbo said. Intrigued, Torquato ran some algorithms on digital images of the retinal patterns and "was astounded," Corbo recalled, "to see the same phenomenon occurring in these systems as they'd seen in a lot of inorganic or physical systems."

Torquato had been studying this hidden order since the early 2000s, when he dubbed it "hyperuniformity." (This term has largely won out over "superhomogeneity," coined around the same time by Joel Lebowitz of Rutgers University.) Since then, it has turned up in a rapidly expanding family of systems. Beyond bird eyes, hyperuniformity is found in materials called quasicrystals, as well as in mathematical matrices full of random numbers, the large-scale structure of the universe, quantum ensembles and soft-matter systems like emulsions and colloids.[1]

Scientists are nearly always taken by surprise when it pops up in new places, as if playing whack-a-mole with the universe. They are still searching for a unifying concept underlying these occurrences. In the process, they've uncovered novel properties of hyperuniform materials that could prove technologically useful.

From a mathematical standpoint, "the more you study it, the more elegant and conceptually compelling it seems," said Henry Cohn, a mathematician and packing expert at Microsoft Research New England, referring to hyperuniformity. "On the other hand, what surprises me about it is the potential breadth of its applications."

A SECRET ORDER

Torquato and a colleague launched the study of hyperuniformity 15 years ago, describing it theoretically and identifying a simple yet surprising example: "You take marbles, you put them in a container, you shake them up until they jam," Torquato said in his Princeton office. "That system is hyperuniform."[2]

The marbles fall into an arrangement, technically called the "maximally random jammed packing," in which they fill 64 percent of space. (The rest is empty air.) This is less than in the densest possible arrangement of spheres—the lattice packing used to stack oranges in a crate, which fills 74 percent of space. But lattice packings aren't always possible to achieve. You can't easily shake a boxful of marbles into a crystalline arrangement. Neither can you form a lattice, Torquato explained, by arranging objects of five different sizes, such as the cones in chicken eyes.

As stand-ins for cones, consider coins on a tabletop. "If you take pennies, and you try to compress the pennies, the pennies like to go into the triangular lattice," Torquato said. But throw some nickels in with the pennies, and "that stops it from crystallizing. Now if you have five different components—throw in quarters, throw in dimes, whatever—that inhibits crystallization even further." Likewise, geometry demands that avian cone cells be disordered. But there's a competing evolutionary demand for the retina to sample light as uniformly as possible, with blue cones positioned far from other blue cones, reds far from other reds and so on. Balancing these constraints, the system "settles for disordered hyperuniformity," Torquato said.

Hyperuniformity gives birds the best of both worlds: Five cone types, arranged in near-uniform mosaics, provide phenomenal color resolution. But it's a "hidden order that you really can't detect with your eye," he said.

Determining whether a system is hyperuniform requires algorithms that work rather like a game of ring toss. First, Torquato said, imagine repeatedly tossing a ring onto an orderly lattice of dots, and each time it lands, counting the number of dots inside the ring. The number of captured dots fluctuates from one ring toss to the next—but not by very much. That's because the interior of the ring always covers a fixed block of dots; the only variation in the number of captured dots happens along the ring's perimeter. If you increase the size of the ring, you will get variation along a longer perimeter. And so with a lattice, the variation in the number of captured dots (or "density fluctuations" in the lattice) grows in proportion to the length of the ring's perimeter. (In higher spatial dimensions, the density fluctuations also scale in proportion to the number of dimensions minus one.)

Now imagine playing ring toss with a smattering of uncorrelated dots—a random distribution, marked by gaps and clusters. A hallmark of randomness is that, as you make the ring bigger, the variation in the number of captured dots scales in proportion to the ring's area, rather than its perimeter. The result is that on large scales, the density fluctuations between ring tosses in a random distribution are much more extreme than in a lattice.

The game gets interesting when it involves hyperuniform distributions. The dots are locally disordered, so for small ring sizes, the number of captured dots fluctuates from one toss to the next more than in a lattice. But as you make the ring bigger, the density fluctuations begin to grow in proportion to the ring's perimeter, rather than its area. This means that the large-scale density of the distribution is just as uniform as that of a lattice.

Among hyperuniform systems, researchers have found a further "zoology of structures," said the Princeton physicist Paul Steinhardt. In these systems, the growth of density fluctuations depends on different powers (between one and two) of the ring's perimeter, multiplied by different coefficients.

"What does it all mean?" Torquato said. "We don't know. It's evolving. There are a lot of papers coming out."

MATERIAL MENAGERIE

Hyperuniformity is clearly a state to which diverse systems converge, but the explanation for its universality is a work in progress. "I see hyperuniformity as basically a hallmark of deeper optimization processes of some sort," Cohn said. But what these processes are "might vary a lot between different problems."

Hyperuniform systems fall into two main classes. Those in the first class, such as quasicrystals—bizarre solids whose interlocked atoms follow no repeating pattern, yet tessellate space—appear to be hyperuniform upon reaching equilibrium, the stable configuration that particles settle into of their own accord. In these equilibrium systems, it is mutual repulsions between the particles that space them apart and give rise to global hyperuniformity. Similar math might explain the emergence of hyperuniformity in bird eyes, the distribution of eigenvalues of random matrices and the zeros of the Riemann zeta function—cousins of the prime numbers.

The other class is not as well understood. In these "nonequilibrium" systems, which include shaken marbles, emulsions, colloids and ensembles

Finding Hidden Order

Imagine repeatedly tossing a ring onto a field of dots and counting the number of whole dots inside the ring each time it lands.

ORDERED LATTICE

The number of enclosed dots per ring toss varies more for large rings than for small rings. This is because all the variation occurs along the ring's edge and so is proportional to the ring's perimeter.

RANDOM DISTRIBUTION

The variation in the number of enclosed dots is proportional to the ring's area, since the density of dots varies throughout the ring. This means the variation can become extreme on large scales.

HYPERUNIFORM DISTRIBUTION

For small rings, the variation is similar to that of a random distribution. But the variation is proportional to the ring's perimeter rather than its area, so for large rings, the variation resembles that of a lattice.

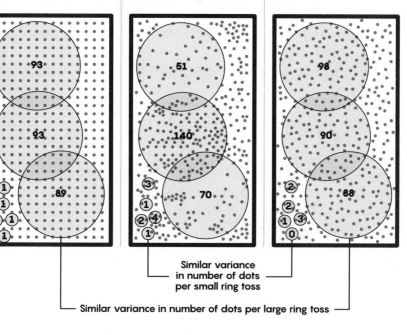

Similar variance
in number of dots
per small ring toss

Similar variance in number of dots per large ring toss

FIGURE 2.4

of cold atoms, particles bump into one another but otherwise do not exert mutual forces; external forces must be applied to the systems to drive them to a hyperuniform state. Within the nonequilibrium class, there are further, intractable divisions. In 2015, physicists led by Denis Bartolo of the École Normale Supérieure in Lyon, France, reported in *Physical Review Letters* that hyperuniformity can be induced in emulsions by sloshing them at the exact amplitude that marks the transition between reversibility and irreversibility in the material: When sloshed more gently than this critical amplitude, the particles suspended in the emulsion return to their previous relative positions after each slosh; when sloshed harder, the particles' motions do not reverse.[3] Bartolo's work suggests a fundamental (though not fully formed) connection between the onset of reversibility and the emergence of hyperuniformity in such nonequilibrium systems. Maximally random jammed packings, meanwhile, are a whole different story.[4] "Can we connect the two physics?" Bartolo said. "No. Not at all. We have absolutely no idea why hyperuniformity shows up in these two very different sets of physical systems."

As they strive to link these threads, scientists have also encountered surprising properties of hyperuniform materials—behaviors that are normally associated with crystals, but which are less susceptible to fabrication errors, more like properties of glass and other uncorrelated disordered media. In a 2016 paper, French physicists led by Rémi Carminati reported that dense hyperuniform materials can be made transparent, whereas uncorrelated disordered materials with the same density would be opaque.[5] The hidden order in the particles' relative positions causes their scattered light to interfere and cancel out. "The interferences destroy scattering," Carminati explained. "Light goes through, as if the material was homogeneous." It's too early to know what dense, transparent, noncrystalline materials might be useful for, Carminati said, but "there are certainly potential applications," particularly in photonics.

And Bartolo's finding about how hyperuniformity is generated in emulsions translates into an easy recipe for stirring concrete, cosmetic creams, glass and food. "Whenever you want to disperse particles inside a paste, you have to deal with a hard mixing problem," he said. "This could be a way to disperse solid particles in a very uniform fashion." First, you identify a material's characteristic amplitude, then you drive it at that amplitude a few dozen times, and an evenly mixed, hyperuniform distribution emerges. "I should not tell you this for free, but rather start a company!" Bartolo said.

Torquato, Steinhardt and associates have already done so. Their start-up, Etaphase, will manufacture hyperuniform photonic circuits—devices that transmit data via light rather than electrons. The Princeton scientists discovered a few years ago that hyperuniform materials can have "band gaps," which block certain frequencies from propagating.[6] Band gaps enable controlled transmission of data, since the blocked frequencies can be contained and guided through channels called waveguides. But band gaps were once thought to be unique to crystal lattices and direction-dependent, aligning with the crystal's symmetry axes. This meant photonic waveguides could only go in certain directions, limiting their use as circuits. Since hyperuniform materials have no preferred direction, their little-understood band gaps are potentially much more practical, enabling not only "wiggly waveguides, but waveguides as you wish," Steinhardt said.

As for the pattern of five-color mosaics in birds' eyes, termed "multihyperuniform," it is, so far, unique in nature. Corbo still hasn't pinpointed how the pattern forms. Does it emerge from mutual repulsions between cone cells, like other systems in the equilibrium class? Or do cones get shaken up like a box of marbles? His guess is the former. Cells can secrete molecules that repel cells of the same type but have no effect on other types; probably, during embryonic development, each cone cell signals that it is differentiating as a certain type, preventing neighboring cells from doing the same. "That's a simple model of how this could develop," he said. "Local action around each cell is creating a global pattern."

Aside from chickens (the most readily available fowl for laboratory study), the same multihyperuniform retinal pattern has turned up in the three other bird species that Corbo has investigated, suggesting that the adaptation is widespread and not tailored to any particular environment. He wonders whether evolution might have found a different optimal configuration in nocturnal species. "That would be super interesting," he said. "It's trickier for us to get our hands on, say, owl eyes."

A UNIFIED THEORY OF RANDOMNESS

Kevin Hartnett

S tandard geometric objects can be described by simple rules—every straight line, for example, is just $y = ax + b$—and they stand in neat relation to each other: Connect two points to make a line, connect four line segments to make a square, connect six squares to make a cube.

These are not the kinds of objects that concern Scott Sheffield. Sheffield, a professor of mathematics at the Massachusetts Institute of Technology, studies shapes that are constructed by random processes. No two of them are ever exactly alike. Consider the most familiar random shape, the random walk, which shows up everywhere from the movement of financial asset prices to the path of particles in quantum physics. These walks are described as random because no knowledge of the path up to a given point can allow you to predict where it will go next.

Beyond the one-dimensional random walk, there are many other kinds of random shapes. There are varieties of random paths, random two-dimensional surfaces, random growth models that approximate, for example, the way a lichen spreads on a rock. All of these shapes emerge naturally in the physical world, yet until recently they've existed beyond the boundaries of rigorous mathematical thought. Given a large collection of random paths or random two-dimensional shapes, mathematicians would have been at a loss to say much about what these random objects shared in common.

Yet in work over the past few years, Sheffield and his frequent collaborator, Jason Miller, a professor at the University of Cambridge, have shown that these random shapes can be categorized into various classes, that these classes have distinct properties of their own and that some kinds of random objects have surprisingly clear connections with other kinds of random objects. Their work forms the beginning of a unified theory of geometric randomness.

"You take the most natural objects—trees, paths, surfaces—and you show they're all related to each other," Sheffield said. "And once you have these relationships, you can prove all sorts of new theorems you couldn't prove before."

Sheffield and Miller published a series of papers that for the first time provides a comprehensive view of random two-dimensional surfaces—an achievement not unlike the Euclidean mapping of the plane.[1]

"Scott and Jason have been able to implement natural ideas and not be rolled over by technical details," said Wendelin Werner, a professor at ETH Zurich and winner of the Fields Medal in 2006 for his work in probability theory and statistical physics. "They have been basically able to push for results that looked out of reach using other approaches."

A RANDOM WALK ON A QUANTUM STRING

In standard Euclidean geometry, objects of interest include lines, rays and smooth curves like circles and parabolas. The coordinate values of the points in these shapes follow clear, ordered patterns that can be described by functions. If you know the value of two points on a line, for instance, you know the values of all other points on the line. The same is true for the values of the points on each of the rays in figure 2.5, which begin at a point and radiate outward.

FIGURE 2.5
Courtesy of Scott Sheffield

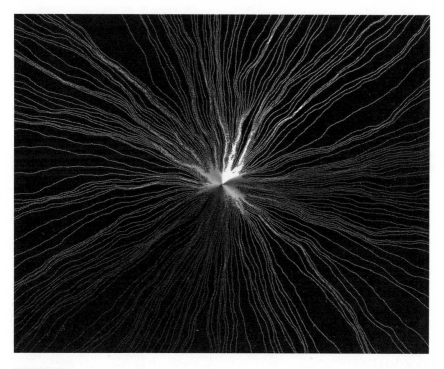

FIGURE 2.6
Courtesy of Scott Sheffield

One way to begin to picture what random two-dimensional geometries look like is to think about airplanes. When an airplane flies a long-distance route, like the route from Tokyo to New York, the pilot flies in a straight line from one city to the other. Yet if you plot the route on a map, the line appears to be curved. The curve is a consequence of mapping a straight line on a sphere (Earth) onto a flat piece of paper.

If Earth were not round, but were instead a more complicated shape, possibly curved in wild and random ways, then an airplane's trajectory (as shown on a flat two-dimensional map) would appear even more irregular, like the rays in figures 2.6–2.8.

Each ray represents the trajectory an airplane would take if it started from the origin and tried to fly as straight as possible over a randomly fluctuating geometric surface. The amount of randomness that characterizes the surface is dialed up in figures 2.6–2.8—as the randomness increases, the straight rays wobble and distort, turn into increasingly jagged bolts of lightning and become nearly incoherent.

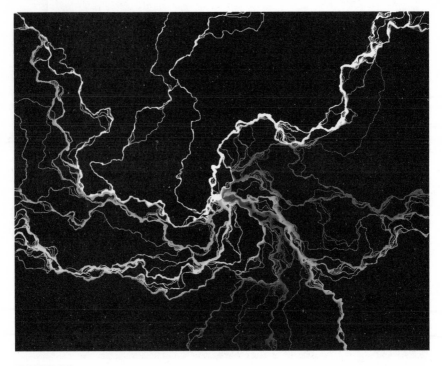

FIGURE 2.7
Courtesy of Scott Sheffield

Yet incoherent is not the same as incomprehensible. In a random geometry, if you know the location of some points, you can (at best) assign probabilities to the location of subsequent points. And just like a loaded set of dice is still random, but random in a different way than a fair set of dice, it's possible to have different probability measures for generating the coordinate values of points on random surfaces.

What mathematicians have found—and hope to continue to find—is that certain probability measures on random geometries are special, and tend to arise in many different contexts. It is as though nature has an inclination to generate its random surfaces using a very particular kind of die (one with an uncountably infinite number of sides). Mathematicians like Sheffield and Miller work to understand the properties of these dice (and the "typical" properties of the shapes they produce) just as precisely as mathematicians understand the ordinary sphere.

The first kind of random shape to be understood in this way was the random walk. Conceptually, a one-dimensional random walk is the kind of path

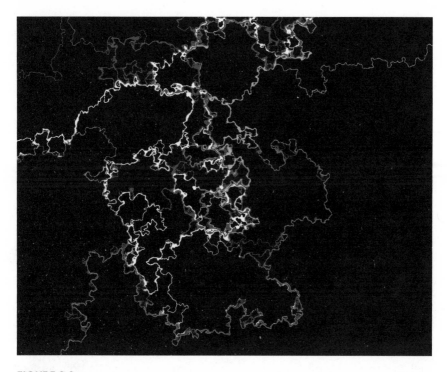

FIGURE 2.8
Courtesy of Scott Sheffield

you'd get if you repeatedly flipped a coin and walked one way for heads and
the other way for tails. In the real world, this type of movement first came
to attention in 1827 when the Scottish botanist Robert Brown observed the
random movements of pollen grains suspended in water. The seemingly random motion was caused by individual water molecules bumping into each
pollen grain. Later, in the 1920s, Norbert Wiener of MIT gave a precise mathematical description of this process, which is called Brownian motion.

Brownian motion is the "scaling limit" of random walks—if you consider a random walk where each step size is very small, and the amount of
time between steps is also very small, these random paths look more and
more like Brownian motion. It's the shape that almost all random walks
converge to over time.

Two-dimensional random spaces, in contrast, first preoccupied physicists as they tried to understand the structure of the universe.

In string theory, one considers tiny strings that wiggle and evolve in time.
Just as the time trajectory of a point can be plotted as a one-dimensional

curve, the time trajectory of a string can be understood as a two-dimensional curve. This curve, called a worldsheet, encodes the history of the one-dimensional string as it wriggles through time.

"To make sense of quantum physics for strings," said Sheffield, "you want to have something like Brownian motion for surfaces."

For years, physicists have had something like that, at least in part. In the 1980s, physicist Alexander Polyakov, who's now at Princeton University, came up with a way of describing these surfaces that came to be called Liouville quantum gravity (LQG).[2] It provided an incomplete but still useful view of random two-dimensional surfaces. In particular, it gave physicists a way of defining a surface's angles so that they could calculate the surface area.

In parallel, another model, called the Brownian map, provided a different way to study random two-dimensional surfaces. Where LQG facilitates calculations about area, the Brownian map has a structure that allows researchers to calculate distances between points. Together, the Brownian map and LQG gave physicists and mathematicians two complementary perspectives on what they hoped were fundamentally the same object. But they couldn't prove that LQG and the Brownian map were in fact compatible with each other.

"It was this weird situation where there were two models for what you'd call the most canonical random surface, two competing random surface models, that came with different information associated with them," said Sheffield.

Beginning in 2013, Sheffield and Miller set out to prove that these two models described fundamentally the same thing.

THE PROBLEM WITH RANDOM GROWTH

Sheffield and Miller began collaborating thanks to a kind of dare. As a graduate student at Stanford in the early 2000s, Sheffield worked under Amir Dembo, a probability theorist. In his dissertation, Sheffield formulated a problem having to do with finding order in a complicated set of surfaces. He posed the question as a thought exercise as much as anything else.

"I thought this would be a problem that would be very hard and take 200 pages to solve and probably nobody would ever do it," Sheffield said.

But along came Miller. In 2006, a few years after Sheffield had graduated, Miller enrolled at Stanford and also started studying under Dembo, who assigned him to work on Sheffield's problem as way of getting to know random processes. "Jason managed to solve this, I was impressed, we started working on some things together, and eventually we had a chance to hire him at MIT as a postdoc," Sheffield said.

In order to show that LQG and the Brownian map were equivalent models of a random two-dimensional surface, Sheffield and Miller adopted an approach that was simple enough conceptually. They decided to see if they could invent a way to measure distance on LQG surfaces and then show that this new distance measurement was the same as the distance measurement that came packaged with the Brownian map.

To do this, Sheffield and Miller thought about devising a mathematical ruler that could be used to measure distance on LQG surfaces. Yet they immediately realized that ordinary rulers would not fit nicely into these random surfaces—the space is so wild that one cannot move a straight object around without the object getting torn apart.

The duo forgot about rulers. Instead, they tried to reinterpret the distance question as a question about growth. To see how this works, imagine a bacterial colony growing on some surface. At first it occupies a single point, but as time goes on it expands in all directions. If you wanted to measure the distance between two points, one (seemingly roundabout) way of doing that would be to start a bacterial colony at one point and measure how much time it took the colony to encompass the other point. Sheffield said that the trick is to somehow "describe this process of gradually growing a ball."

It's easy to describe how a ball grows in the ordinary plane, where all points are known and fixed and growth is deterministic. Random growth is far harder to describe and has long vexed mathematicians. Yet as Sheffield and Miller were soon to learn, "[random growth] becomes easier to understand on a random surface than on a smooth surface," said Sheffield. The randomness in the growth model speaks, in a sense, the same language as the randomness on the surface on which the growth model proceeds. "You add a crazy growth model on a crazy surface, but somehow in some ways it actually makes your life better," he said.

Figures 2.9–2.11 show a specific random growth model, the Eden model, which describes the random growth of bacterial colonies. The colonies grow through the addition of randomly placed clusters along their boundaries. At any given point in time, it's impossible to know for sure where on the boundary the next cluster will appear. In these images, Miller and Sheffield show how Eden growth proceeds over a random two-dimensional surface.

Figure 2.9 shows Eden growth on a fairly flat—that is, not especially random—LQG surface. The growth proceeds in an orderly way, forming nearly

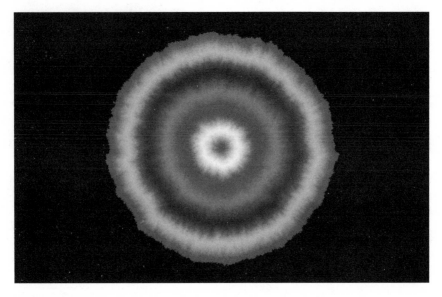

FIGURE 2.9
Eden growth with gamma equal to 0.25. Courtesy of Jason Miller.

concentric circles that have been shaded to indicate the time at which
growth occurs at different points on the surface.

In subsequent images, Sheffield and Miller illustrate growth on surfaces
of increasingly greater randomness. The amount of randomness in the
function that produces the surfaces is controlled by a constant, gamma.
As gamma increases, the surface gets rougher—with higher peaks and
lower valleys—and random growth on that surface similarly takes on a less
orderly form. In figure 2.9, gamma is 0.25. In figure 2.10, gamma is set to
1.25, introducing five times as much randomness into the construction of
the surface. Eden growth across this uncertain surface is similarly distorted.

When gamma is set to the square root of eight-thirds (approximately
1.63; see figure 2.11), LQG surfaces fluctuate even more dramatically. They
also take on a roughness that matches the roughness of the Brownian map,
which allows for more direct comparisons between these two models of a
random geometric surface.

Random growth on such a rough surface proceeds in a very irregular
way. Describing it mathematically is like trying to anticipate minute pressure fluctuations in a hurricane. Yet Sheffield and Miller realized that they
needed to figure out how to model Eden growth on very random LQG surfaces in order to establish a distance structure equivalent to the one on the
(very random) Brownian map.

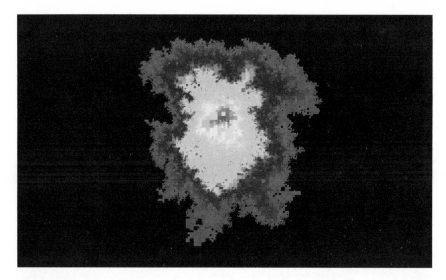

FIGURE 2.10
Eden growth with gamma equal to 1.25. Courtesy of Jason Miller.

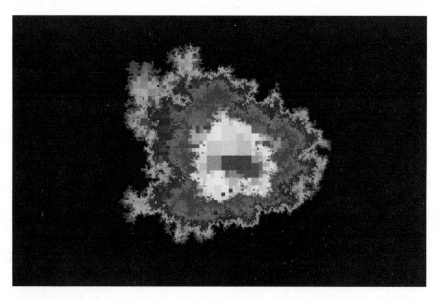

FIGURE 2.11
Eden growth with gamma equal to 1.63. Courtesy of Jason Miller.

"Figuring out how to mathematically make [random growth] rigorous is a huge stumbling block," said Sheffield, noting that Martin Hairer, now of Imperial College London, won the Fields Medal in 2014 for work that overcame just these kinds of obstacles. "You always need some kind of amazing clever trick to do it."

RANDOM EXPLORATION

Sheffield and Miller's clever trick is based on a special type of random one-dimensional curve that is similar to the random walk except that it never crosses itself. Physicists had encountered these kinds of curves for a long time in situations where, for instance, they were studying the boundary between clusters of particles with positive and negative spin (the boundary line between the clusters of particles is a one-dimensional path that never crosses itself and takes shape randomly). They knew these kinds of random, noncrossing paths occurred in nature, just as Robert Brown had observed that random crossing paths occurred in nature, but they didn't know how to think about them in any kind of precise way. In 1999 Oded Schramm, who at the time was at Microsoft Research in Redmond, Washington, introduced the SLE curve (for Schramm–Loewner evolution) as the canonical noncrossing random curve.

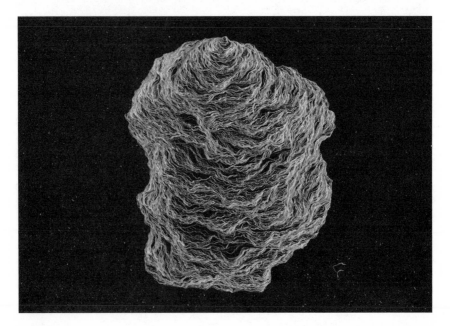

FIGURE 2.12
An example of an SLE curve. Courtesy of Jason Miller.

Schramm's work on SLE curves was a landmark in the study of random objects. It's widely acknowledged that Schramm, who died in a hiking accident in 2008, would have won the Fields Medal had he been a few weeks younger at the time he'd published his results. (The Fields Medal can be given only to mathematicians who are not yet 40.) As it was, two people who worked with him built on his work and went on to win the prize: Wendelin Werner in 2006 and Stanislav Smirnov in 2010. More fundamentally, the discovery of SLE curves made it possible to prove many other things about random objects.

"As a result of Schramm's work, there were a lot of things in physics they'd known to be true in their physics way that suddenly entered the realm of things we could prove mathematically," said Sheffield, who was a friend and collaborator of Schramm's.

For Miller and Sheffield, SLE curves turned out to be valuable in an unexpected way. In order to measure distance on LQG surfaces, and thus show that LQG surfaces and the Brownian map were the same, they needed to find some way to model random growth on a random surface. SLE proved to be the way.

"The 'aha' moment was [when we realized] you can construct [random growth] using SLEs and that there is a connection between SLEs and LQG," said Miller.

SLE curves come with a constant, kappa, which plays a similar role to the one gamma plays for LQG surfaces. Where gamma describes the roughness of an LQG surface, kappa describes the "windiness" of SLE curves. When kappa is low, the curves look like straight lines. As kappa increases, more randomness is introduced into the function that constructs the curves and the curves turn more unruly, while obeying the rule that they can bounce off of, but never cross, themselves. Figure 2.13 shows an SLE curve with kappa equal to 0.5, followed by an SLE curve in figure 2.14 with kappa equal to 3.

Sheffield and Miller noticed that when they dialed the value of kappa to 6 and gamma up to the square root of eight-thirds, an SLE curve drawn on the random surface followed a kind of exploration process. Thanks to works by Schramm and by Smirnov, Sheffield and Miller knew that when kappa equals 6, SLE curves follow the trajectory of a kind of "blind explorer" who marks her path by constructing a trail as she goes. She moves as randomly as possible except that whenever she bumps into a piece of the path she has already followed, she turns away from that piece to avoid crossing her own path or getting stuck in a dead end.

"[The explorer] finds that each time her path hits itself, it cuts off a little piece of land that is completely surrounded by the path and can never be visited again," said Sheffield.

FIGURE 2.13
An SLE curve with kappa equal to 0.5. Courtesy of Scott Sheffield.

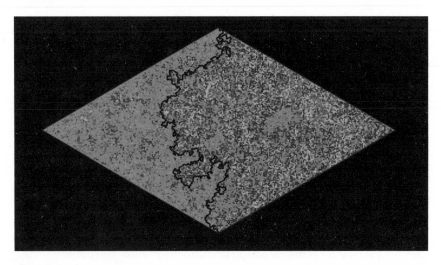

FIGURE 2.14
An SLE curve with kappa equal to 3. Courtesy of Scott Sheffield.

Sheffield and Miller then considered a bacterial growth model, the Eden model, that had a similar effect as it advanced across a random surface: It grew in a way that "pinched off" a plot of terrain that, afterward, it never visited again. The plots of terrain cut off by the growing bacteria colony looked exactly the same as the plots of terrain cut off by the blind explorer.

Moreover, the information possessed by a blind explorer at any time about the outer unexplored region of the random surface was exactly the same as the information possessed by a bacterial colony. The only difference between the two was that while the bacterial colony grew from all points on its outer boundary at once, the blind explorer's SLE path could grow only from the tip.

In a paper posted online in 2013, Sheffield and Miller imagined what would happen if, every few minutes, the blind explorer were magically transported to a random new location on the boundary of the territory she had already visited. By moving all around the boundary, she would be effectively growing her path from all boundary points at once, much like the bacterial colony. Thus they were able to take something they could understand—how an SLE curve proceeds on a random surface—and show that with some special configuring, the curve's evolution exactly described a process they hadn't been able to understand: random growth. "There's something special about the relationship between SLE and growth," said Sheffield. "That was kind of the miracle that made everything possible."

The distance structure imposed on LQG surfaces through the precise understanding of how random growth behaves on those surfaces exactly matched the distance structure on the Brownian map. As a result, Sheffield and Miller merged two distinct models of random two-dimensional shapes into one coherent, mathematically understood fundamental object.

TURNING RANDOMNESS INTO A TOOL

Sheffield and Miller have already posted the first two papers in their proof of the equivalence between LQG and the Brownian map on the scientific preprint site arxiv.org.[3] The work turned on the ability to reason across different random shapes and processes—to see how random noncrossing curves, random growth and random two-dimensional surfaces relate to one another. It's an example of the increasingly sophisticated results that are possible in the study of random geometry.

"It's like you're in a mountain with three different caves. One has iron, one has gold, one has copper—suddenly you find a way to link all three of these caves together," said Sheffield. "Now you have all these different elements you can build things with and can combine them to produce all sorts of things you couldn't build before."

Many open questions remain, including determining whether the relationship between SLE curves, random growth models and distance measurements holds up in less-rough versions of LQG surfaces than the one used in the recent paper. In practical terms, the results by Sheffield and Miller

can be used to describe the random growth of real phenomena like snow-
flakes, mineral deposits and dendrites in caves, but only when that growth
takes place in the imagined world of random surfaces. It remains to be seen
whether their methods can be applied to ordinary Euclidean space, like the
space we live in.

STRANGE NUMBERS FOUND IN PARTICLE COLLISIONS

Kevin Hartnett

At the Large Hadron Collider in Geneva, physicists shoot protons around a 17-mile track and smash them together at nearly the speed of light. It's one of the most finely tuned scientific experiments in the world, but when trying to make sense of the quantum debris, physicists begin with a strikingly simple tool called a Feynman diagram that's not that different from how a child would depict the situation.

Feynman diagrams were devised by Richard Feynman in the 1940s. They feature lines representing elementary particles that converge at a vertex (which represents a collision) and then diverge from there to represent the pieces that emerge from the crash. Those lines either shoot off alone or converge again. The chain of collisions can be as long as a physicist dares to consider.

To that schematic physicists then add numbers, for the mass, momentum and direction of the particles involved. Then they begin a laborious accounting procedure—integrate these, add that, square this. The final result is a single number, called a Feynman probability, which quantifies the chance that the particle collision will play out as sketched.

"In some sense Feynman invented this diagram to encode complicated math as a bookkeeping device," said Sergei Gukov, a theoretical physicist and mathematician at Caltech.

Feynman diagrams have served physics well over the years, but they have limitations. One is strictly procedural. Physicists are pursuing increasingly high-energy particle collisions that require greater precision of measurement—and as the precision goes up, so does the intricacy of the Feynman diagrams that need to be calculated to generate a prediction.

The second limitation is of a more fundamental nature. Feynman diagrams are based on the assumption that the more potential collisions and sub-collisions physicists account for, the more accurate their numerical predictions will be. This process of calculation, known as perturbative expansion, works very well for particle collisions of electrons, where the weak and electromagnetic forces dominate. It works less well for high-energy collisions, like collisions between protons, where the strong nuclear force prevails. In these cases, accounting for a wider range of collisions—by drawing ever more elaborate Feynman diagrams—can actually lead physicists astray.

"We know for a fact that at some point it begins to diverge" from real-world physics, said Francis Brown, a mathematician at the University of Oxford. "What's not known is how to estimate at what point one should stop calculating diagrams."

Yet there is reason for optimism. Over the last decade physicists and mathematicians have been exploring a surprising correspondence that has the potential to breathe new life into the venerable Feynman diagram and generate far-reaching insights in both fields. It has to do with the strange fact that the values calculated from Feynman diagrams seem to exactly match some of the most important numbers that crop up in a branch of mathematics known as algebraic geometry. These values are called "periods of motives," and there's no obvious reason why the same numbers should appear in both settings. Indeed, it's as strange as it would be if every time you measured a cup of rice, you observed that the number of grains was prime.

"There is a connection from nature to algebraic geometry and periods, and with hindsight, it's not a coincidence," said Dirk Kreimer, a physicist at Humboldt University in Berlin.

Now mathematicians and physicists are working together to unravel the coincidence. For mathematicians, physics has called to their attention a special class of numbers that they'd like to understand: Is there a hidden structure to these periods that occur in physics? What special properties might this class of numbers have? For physicists, the reward of that kind of mathematical understanding would be a new degree of foresight when it comes to anticipating how events will play out in the messy quantum world.

A RECURRING THEME

Today, periods are one of the most abstract subjects of mathematics, but they started out as a more concrete concern. In the early 17th century scientists such as Galileo Galilei were interested in figuring out how to calculate the length of time a pendulum takes to complete a swing. They realized that the calculation boiled down to taking the integral—a kind of infinite sum—of a function that combined information about the pendulum's length and angle of release. Around the same time, Johannes Kepler used similar calculations to establish the time that a planet takes to travel around the sun. They called these measurements "periods" and established them as one of the most important measurements that can be made about motion.

Over the course of the 18th and 19th centuries, mathematicians became interested in studying periods generally—not just as they related to pendulums or planets, but as a class of numbers generated by integrating polynomial functions like $x^2 + 2x - 6$ and $3x^3 - 4x^2 - 2x + 6$. For more than a century, luminaries like Carl Friedrich Gauss and Leonhard Euler explored the universe of periods and found that it contained many features that pointed to some underlying order. In a sense, the field of algebraic geometry—which studies the geometric forms of polynomial equations—developed in the 20th century as a means for pursuing that hidden structure.

This effort advanced rapidly in the 1960s. By that time mathematicians had done what they often do: They translated relatively concrete objects like equations into more abstract ones, which they hoped would allow them to identify relationships that were not initially apparent.

This process first involved looking at the geometric objects (known as algebraic varieties) defined by the solutions to classes of polynomial functions, rather than looking at the functions themselves. Next, mathematicians tried to understand the basic properties of those geometric objects. To do that they developed what are known as cohomology theories—ways of identifying structural aspects of the geometric objects that were the same regardless of the particular polynomial equation used to generate the objects.

By the 1960s, cohomology theories had proliferated to the point of distraction—singular cohomology, de Rham cohomology, étale cohomology and so on. Everyone, it seemed, had a different view of the most important features of algebraic varieties.

It was in this cluttered landscape that the pioneering mathematician Alexander Grothendieck, who died in 2014, realized that all cohomology theories were different versions of the same thing.

"What Grothendieck observed is that, in the case of an algebraic variety, no matter how you compute these different cohomology theories, you always somehow find the same answer," Brown said.

That same answer—the unique thing at the center of all these cohomology theories—was what Grothendieck called a "motive." "In music it means a recurring theme. For Grothendieck a motive was something which is coming again and again in different forms, but it's really the same," said Pierre Cartier, a mathematician at the Institute of Advanced Scientific Studies outside Paris and a former colleague of Grothendieck's.

Motives are in a sense the fundamental building blocks of polynomial equations, in the same way that prime factors are the elemental pieces of larger numbers. Motives also have their own data associated with them. Just as you can break matter into elements and specify characteristics of each element—its atomic number and atomic weight and so forth—mathematicians ascribe essential measurements to a motive. The most important of these measurements are the motive's periods. And if the period of a motive arising in one system of polynomial equations is the same as the period of a motive arising in a different system, you know the motives are the same.

"Once you know the periods, which are specific numbers, that's almost the same as knowing the motive itself," said Minhyong Kim, a mathematician at Oxford.

One direct way to see how the same period can show up in unexpected contexts is with pi, "the most famous example of getting a period," Cartier said. Pi shows up in many guises in geometry: in the integral of the function that defines the one-dimensional circle, in the integral of the function that defines the two-dimensional circle and in the integral of the function that defines the sphere. That this same value would recur in such seemingly different-looking integrals was likely mysterious to ancient thinkers. "The modern explanation is that the sphere and the solid circle have the same motive and therefore have to have essentially the same period," Brown wrote in an email.

FEYNMAN'S ARDUOUS PATH

If curious minds long ago wanted to know why values like pi crop up in calculations on the circle and the sphere, today mathematicians and physicists would like to know why those values arise out of a different kind of geometric object: Feynman diagrams.

Feynman diagrams have a basic geometric aspect to them, formed as they are from line segments, rays and vertices. To see how they're constructed, and why they're useful in physics, imagine a simple experimental setup in which an electron and a positron collide to produce a muon and an antimuon. To calculate the probability of that result taking place, a physicist would need to know the mass and momentum of each of the incoming particles and also something about the path the particles followed. In quantum mechanics, the path a particle takes can be thought of as the average of all the possible paths it might take. Computing that path becomes a matter of taking an integral, known as a Feynman path integral, over the set of all paths.

Every route a particle collision could follow from beginning to end can be represented by a Feynman diagram, and each diagram has its own associated integral. (The diagram and its integral are one and the same.) To calculate the probability of a specific outcome from a specific set of starting conditions, you consider all possible diagrams that could describe what happens, take each integral and add those integrals together. That number is the diagram's amplitude. Physicists then square the magnitude of this number to get the probability.

This procedure is easy to execute for an electron and a positron going in and a muon and an antimuon coming out. But that's boring physics. The experiments that physicists really care about involve Feynman diagrams with loops. Loops represent situations in which particles emit and then reabsorb additional particles. When an electron collides with a positron, there's an infinite number of intermediate collisions that can take place before the final muon and antimuon pair emerges. In these intermediate collisions, new particles like photons are created and annihilated before they can be observed. The entering and exiting particles are the same as previously described, but the fact that those unobservable collisions happen can still have subtle effects on the outcome.

"It's like Tinkertoys. Once you draw a diagram you can connect more lines according to the rules of the theory," said Flip Tanedo, a physicist at the University of California, Riverside. "You can connect more sticks, more nodes, to make it more complicated."

By considering loops, physicists increase the precision of their experiments. (Adding a loop is like calculating a value to a greater number of significant digits). But each time they add a loop, the number of Feynman diagrams that need to be considered—and the difficulty of the corresponding integrals—goes up dramatically. For example, a one-loop version of a simple system might require just one diagram. A two-loop version of the same system needs seven diagrams. Three loops demand 72 diagrams. Increase it to five loops, and the calculation requires around 12,000 integrals—a computational load that can literally take years to resolve.

Rather than chugging through so many tedious integrals, physicists would love to gain a sense of the final amplitude just by looking at the structure of a given Feynman diagram—just as mathematicians can associate periods with motives.

"This procedure is so complex and the integrals are so hard, so what we'd like to do is gain insight about the final answer, the final integral or period, just by staring at the graph," Brown said.

A SURPRISING CONNECTION

Periods and amplitudes were presented together for the first time in 1994 by Kreimer and David Broadhurst, a physicist at the Open University in England, with a paper following in 1995.[1] The work led mathematicians to speculate that all amplitudes were periods of mixed Tate motives—a special kind of motive named after John Tate, emeritus professor at Harvard University, in which all the periods are multiple values of one of the most influential constructions in number theory, the Riemann zeta function. In the situation with an electron-positron pair going in and a muon-antimuon pair coming out, the main part of the amplitude comes out as six times the Riemann zeta function evaluated at three.

If all amplitudes were multiple zeta values, it would give physicists a well-defined class of numbers to work with. But in 2012 Brown and his collaborator Oliver Schnetz proved that's not the case.[2] While all the amplitudes physicists come across today may be periods of mixed Tate motives, "there are monsters lurking out there that throw a spanner into the works," Brown said. Those monsters are "certainly periods, but they're not the nice and simple periods people had hoped for."

What physicists and mathematicians do know is that there seems to be a connection between the number of loops in a Feynman diagram and a notion in mathematics called "weight." Weight is a number related to the dimension of the space being integrated over: A period integral over a one-dimensional space can have a weight of 0, 1 or 2; a period integral over a two-dimensional space can have weight up to 4, and so on. Weight can also be used to sort periods into different types: All periods of weight 0 are conjectured to be algebraic numbers, which can be the solutions to polynomial equations (this has not been proved); the period of a pendulum always has a weight of 1; pi is a period of weight 2; and the weights of values of the Riemann zeta function are always twice the input (so the zeta function evaluated at 3 has a weight of 6).

This classification of periods by weights carries over to Feynman diagrams, where the number of loops in a diagram is somehow related to the weight of its amplitude. Diagrams with no loops have amplitudes of weight 0; the amplitudes of diagrams with one loop are all periods of mixed Tate motives and have, at most, a weight of 4. For graphs with additional loops, mathematicians suspect the relationship continues, even if they can't see it yet.

"We go to higher loops and we see periods of a more general type," Kreimer said. "There mathematicians get really interested because they don't understand much about motives that are not mixed Tate motives."

Mathematicians and physicists are currently going back and forth trying to establish the scope of the problem and craft solutions. Mathematicians suggest functions (and their integrals) to physicists that can be used to describe Feynman diagrams. Physicists produce configurations of particle collisions that outstrip the functions mathematicians have to offer. "It's quite amazing to see how fast they've assimilated quite technical mathematical ideas," Brown said. "We've run out of classical numbers and functions to give to physicists."

NATURE'S GROUPS

Since the development of calculus in the 17th century, numbers arising in the physical world have informed mathematical progress. Such is the case today. The fact that the periods that come from physics are "somehow God-given and come from physical theories means they have a lot of structure and it's structure a mathematician wouldn't necessarily think of or try to invent," said Brown.

Kreimer added, "It seems so that the periods which nature wants are a smaller set than the periods mathematics can define, but we cannot define very cleanly what this subset really is."

Brown is looking to prove that there's a kind of mathematical group—a Galois group—acting on the set of periods that come from Feynman diagrams. "The answer seems to be yes in every single case that's ever been computed," he said, but proof that the relationship holds categorically is still in the distance. "If it were true that there were a group acting on the numbers coming from physics, that means you're finding a huge class of symmetries," Brown said. "If that's true, then the next step is to ask why there's this big symmetry group and what possible physics meaning could it have."

Among other things, it would deepen the already provocative relationship between fundamental geometric constructions from two very different contexts: motives, the objects that mathematicians devised 50 years ago to understand the solutions to polynomial equations, and Feynman diagrams, the schematic representation of how particle collisions play out. Every Feynman diagram has a motive attached to it, but what exactly the structure of a motive is saying about the structure of its related diagram remains anyone's guess.

QUANTUM QUESTIONS INSPIRE NEW MATH

Robbert Dijkgraaf

M athematics might be more of an environmental science than we realize. Even though it is a search for eternal truths, many mathematical concepts trace their origins to everyday experience. Astrology and architecture inspired Egyptians and Babylonians to develop geometry. The study of mechanics during the scientific revolution of the 17th century brought us calculus.

Remarkably, ideas from quantum theory turn out to carry tremendous mathematical power as well, even though we have little daily experience dealing with elementary particles. The bizarre world of quantum theory—where things can seem to be in two places at the same time and are subject to the laws of probability—not only represents a more fundamental description of nature than what preceded it, it also provides a rich context for modern mathematics. Could the logical structure of quantum theory, once fully understood and absorbed, inspire a new realm of mathematics that might be called "quantum mathematics"?

There is of course a long-standing and intimate relationship between mathematics and physics. Galileo famously wrote about a book of nature waiting to be decoded: "Philosophy is written in this grand book, the universe, which stands continually open to our gaze. But the book cannot be understood unless one first learns to comprehend the language and read the letters in which it is composed. It is written in the language of mathematics." From more modern times we can quote Richard Feynman, who was not known as a connoisseur of abstract mathematics: "To those who do not know mathematics it is difficult to get across a real feeling as to the beauty, the deepest beauty, of nature. ... If you want to learn about nature, to appreciate nature, it is necessary to understand the language that she speaks in." (On the other hand, he also stated: "If all mathematics disappeared today, physics would be set back exactly one week," to which a mathematician had the clever riposte: "This was the week that God created the world.")

The mathematical physicist and Nobel laureate Eugene Wigner has written eloquently about the amazing ability of mathematics to describe reality, characterizing it as "the unreasonable effectiveness of mathematics in the natural sciences." The same mathematical concepts turn up in a wide range of contexts. But these days we seem to be witnessing the reverse: the unreasonable effectiveness of quantum theory in modern mathematics. Ideas that originate in particle physics have an uncanny tendency to appear in the most diverse mathematical fields. This is especially true for string theory. Its stimulating influence in mathematics will have a lasting and rewarding impact, whatever its final role in fundamental physics turns out to be. The number of disciplines that it touches is dizzying: analysis, geometry, algebra, topology, representation theory, combinatorics, probability—the list goes on and on. One starts to feel sorry for the poor students who have to learn all this!

What could be the underlying reason for this unreasonable effectiveness of quantum theory? In my view, it is closely connected to the fact that in the quantum world everything that can happen does happen.

In a very schematic way, classical mechanics tries to compute how a particle travels from A to B. For example, the preferred path could be along a geodesic—a path of minimal length in a curved space. In quantum mechanics one considers instead the collection of all possible paths from A to B, however long and convoluted. This is Feynman's famous "sum over histories" interpretation. The laws of physics will then assign to each path a certain weight that determines the probability that a particle will move along that particular trajectory. The classical solution that obeys Newton's laws is simply the most likely one among many. So, in a natural way, quantum physics studies the set of all paths, as a weighted ensemble, allowing us to sum over all possibilities.

This holistic approach of considering everything at once is very much in the spirit of modern mathematics, where the study of "categories" of objects focuses much more on the mutual relations than on any specific individual example. It is this bird's-eye view of quantum theory that brings out surprising new connections.

QUANTUM CALCULATORS

A striking example of the magic of quantum theory is mirror symmetry—a truly astonishing equivalence of spaces that has revolutionized geometry.

The story starts in enumerative geometry, a well-established, but not very exciting branch of algebraic geometry that counts objects. For example, researchers might want to count the number of curves on Calabi–Yau spaces—six-dimensional solutions of Einstein's equations of gravity that are of particular interest in string theory, where they are used to curl up extra space dimensions.

Just as you can wrap a rubber band around a cylinder multiple times, the curves on a Calabi–Yau space are classified by an integer, called the degree, that measures how often they wrap around. Finding the numbers of curves of a given degree is a famously hard problem, even for the simplest Calabi–Yau space, the so-called quintic. A classical result from the 19th century states that the number of lines—degree-one curves—is equal to 2,875. The number of degree-two curves was only computed around 1980 and turns out to be much larger: 609,250. But the number of curves of degree three required the help of string theorists.

Around 1990, a group of string theorists asked geometers to calculate this number. The geometers devised a complicated computer program and came back with an answer. But the string theorists suspected it was erroneous, which suggested a mistake in the code. Upon checking, the geometers confirmed there was, but how did the physicists know?

String theorists had already been working to translate this geometric problem into a physical one. In doing so, they had developed a way to calculate the number of curves of any degree all at once.[1] It's hard to overestimate the shock of this result in mathematical circles. It was a bit like devising a way to climb each and every mountain, no matter how high!

Within quantum theory it makes perfect sense to combine the numbers of curves of all degrees into a single elegant function. Assembled in this way, it has a straightforward physical interpretation. It can be seen as a probability amplitude for a string propagating in the Calabi–Yau space, where the sum-over-histories principle has been applied. A string can be thought to probe all possible curves of every possible degree at the same time and is thus a super-efficient "quantum calculator."

But a second ingredient was necessary to find the actual solution: an equivalent formulation of the physics using a so-called "mirror" Calabi–Yau space. The term "mirror" is deceptively simple. In contrast to the way an ordinary mirror reflects an image, here the original space and its mirror are of very different shapes; they do not even have the same topology. But in the realm of quantum theory, they share many properties. In particular, the string propagation in both spaces turns out to be identical. The difficult computation on the original manifold translates into a much simpler expression on the mirror manifold, where it can be computed by a single integral. *Et voilà!*

DUALITY OF EQUALS

Mirror symmetry illustrates a powerful property of quantum theory called duality: Two classical models can become equivalent when considered as quantum systems, as if a magic wand is waved and all the differences suddenly disappear. Dualities point to deep but often mysterious symmetries of the underlying quantum theory. In general, they are poorly understood and an indication that our understanding of quantum theory is incomplete at best.

The first and most famous example of such an equivalence is the well-known particle-wave duality that states that every quantum particle, such as an electron, can be considered both as a particle and as a wave. Both points of views have their advantages, offering different perspectives on the same physical phenomenon. The "correct" point of view—particle or wave—is determined solely by the nature of the question, not by the nature of the electron. The two sides of mirror symmetry offer dual and equally valid perspectives on "quantum geometry."

Mathematics has the wonderful ability to connect different worlds. The most overlooked symbol in any equation is the humble equal sign. Ideas flow through it, as if the equal sign conducts the electric current that illuminates the "Aha!" lightbulb in our mind. And the double lines indicate that ideas can flow in both directions. Albert Einstein was an absolute master of finding equations that exemplify this property. Take $E = mc^2$, without a doubt the most famous equation in history. In all its understated elegance, it connects the physical concepts of mass and energy that were seen as totally distinct before the advent of relativity. Through Einstein's equation we learn that mass can be transformed into energy, and vice versa. The equation of Einstein's general theory of relativity, although less catchy and well-known, links the worlds of geometry and matter in an equally surprising and beautiful manner. A succinct way to summarize that theory is that mass tells space how to curve, and space tells mass how to move.

Mirror symmetry is another perfect example of the power of the equal sign. It is capable of connecting two different mathematical worlds. One is the realm of symplectic geometry, the branch of mathematics that underlies much of mechanics. On the other side is the realm of algebraic geometry, the world of complex numbers. Quantum physics allows ideas to flow freely from one field to the other and provides an unexpected "grand unification" of these two mathematical disciplines.

It is comforting to see how mathematics has been able to absorb so much of the intuitive, often imprecise reasoning of quantum physics and string theory, and to transform many of these ideas into rigorous statements and proofs. Mathematicians are close to applying this exactitude to homological mirror symmetry, a program that vastly extends string theory's original idea of mirror symmetry. In a sense, they're writing a full dictionary of the objects that appear in the two separate mathematical worlds, including all the relations they satisfy. Remarkably, these proofs often do not follow the path that physical arguments had suggested. It is apparently not the role of mathematicians to clean up after physicists! On the contrary, in many cases completely new lines of thought had to be developed in order to find the proofs. This is further evidence of the deep and as yet undiscovered logic that underlies quantum theory and, ultimately, reality.

Niels Bohr was very fond of the notion of complementarity. The concept emerged from the fact that, as Werner Heisenberg proved with his uncertainty principle, in quantum mechanics one can measure either the momentum p of a particle or its position q, but not both at the same time. Wolfgang Pauli wittily summarized this duality in a letter to Heisenberg dated October 19, 1926, just a few weeks after the discovery: "One can see the world with the p-eye, and one can see it with the q-eye, but if one opens both eyes, then one becomes crazy."

In his later years, Bohr tried to push this idea into a much broader philosophy. One of his favorite complementary pairs was truth and clarity. Perhaps the pair of mathematical rigor and physical intuition should be added as another example of two mutually exclusive qualities. You can look at the world with a mathematical eye or with a complementary physical eye, but don't dare to open both.

III HOW ARE SURPRISING PROOFS DISCOVERED?

A PATH LESS TAKEN TO THE PEAK OF THE MATH WORLD

Kevin Hartnett

On a warm spring morning in 2017, June Huh walked across the campus of Princeton University. His destination was McDonnell Hall, where he was scheduled to teach, and he wasn't quite sure how to get there. Huh is a member of the rarefied Institute for Advanced Study, which lies adjacent to Princeton's campus. As a member of IAS, Huh has no obligation to teach, but he'd volunteered to give an advanced undergraduate math course on a topic called commutative algebra. When I asked him why, he replied, "When you teach, you do something useful. When you do research, most days you don't."

We arrived at Huh's classroom a few minutes before class was scheduled to begin. Inside, nine students sat in loose rows. One slept with his head down on the table. Huh took a position in a front corner of the room and removed several pages of crumpled notes from his backpack. Then, with no fanfare, he picked up where he'd left off the previous week. Over the next 80 minutes he walked students through a proof of a theorem by the German mathematician David Hilbert that stands as one of the most important breakthroughs in 20th-century mathematics.

Commutative algebra is taught at the undergraduate level at only a few universities, but it is offered routinely at Princeton, which each year enrolls a handful of the most promising young math minds in the world. Even by that standard, Huh says the students in his class that morning were unusually talented. One of them, sitting that morning in the front row, is the only person ever to have won five consecutive gold medals at the International Mathematical Olympiad.

Huh's math career began with much less acclaim. A bad score on an elementary school test convinced him that he was not very good at math. As a teenager he dreamed of becoming a poet. He didn't major in math, and when he finally applied to graduate school, he was rejected by every university save one.

Nine years later, at the age of 34, Huh is at the pinnacle of the math world. He is best known for his proof, with the mathematicians Eric Katz and Karim Adiprasito, of a long-standing problem called Rota's conjecture.[1]

Even more remarkable than the proof itself is the manner in which Huh and his collaborators achieved it—by finding a way to reinterpret ideas from one area of mathematics in another where they didn't seem to belong. In the spring of 2017, IAS offered Huh a long-term fellowship, a position that has been extended to only three young mathematicians before. Two of them (Vladimir Voevodsky and Ngô Bảo Châu) went on to win the Fields Medal, the highest honor in mathematics.

That Huh would achieve this status after starting mathematics so late is almost as improbable as if he had picked up a tennis racket at 18 and won Wimbledon at 20. It's the kind of out-of-nowhere journey that simply doesn't happen in mathematics today, where it usually takes years of specialized training even to be in a position to make new discoveries. Yet it would be a mistake to see Huh's breakthroughs as having come in spite of his unorthodox beginning. In many ways they're a product of his unique history—a direct result of his chance encounter, in his last year of college, with a legendary mathematician who somehow recognized a gift in Huh that Huh had never perceived himself.

THE ACCIDENTAL APPRENTICE

Huh was born in 1983 in California, where his parents were attending graduate school. They moved back to Seoul, South Korea, when he was 2. There, his father taught statistics and his mother became one of the first professors of Russian literature in South Korea since the onset of the Cold War.

After that bad math test in elementary school, Huh says he adopted a defensive attitude toward the subject: He didn't think he was good at math, so he decided to regard it as a barren pursuit of one logically necessary statement piled atop another. As a teenager he took to poetry instead, viewing it as a realm of true creative expression. "I knew I was smart, but I couldn't demonstrate that with my grades, so I started to write poetry," Huh said.

Huh wrote many poems and a couple of novellas, mostly about his own experiences as a teenager. None were ever published. By the time he enrolled at Seoul National University in 2002, he had concluded that he couldn't make a living as a poet, so he decided to become a science journalist instead. He majored in astronomy and physics, in perhaps an unconscious nod to his latent analytic abilities.

When Huh was 24 and in his last year of college, the famed Japanese mathematician Heisuke Hironaka came to Seoul National as a visiting professor. Hironaka was in his mid-70s at the time and was a full-fledged celebrity in Japan and South Korea. He'd won the Fields Medal in 1970 and later wrote a best-selling memoir called *The Joy of Learning*, which a generation of Korean and Japanese parents had given their kids in the hope of nurturing the next great mathematician. At Seoul National, he taught a yearlong lecture course in a broad area of mathematics called algebraic geometry. Huh attended, thinking Hironaka might become his first subject as a journalist.

Initially Huh was among more than 100 students, including many math majors, but within a few weeks enrollment had dwindled to a handful. Huh imagines other students quit because they found Hironaka's lectures incomprehensible. He says he persisted because he had different expectations about what he might get out of the course.

"The math students dropped out because they could not understand anything. Of course, I didn't understand anything either, but non-math students have a different standard of what it means to understand something," Huh said. "I did understand some of the simple examples he showed in classes, and that was good enough for me."

After class Huh would make a point of talking to Hironaka, and the two soon began having lunch together. Hironaka remembers Huh's initiative. "I didn't reject students, but I didn't always look for students, and he was just coming to me," Hironaka recalled.

Huh tried to use these lunches to ask Hironaka questions about himself, but the conversation kept coming back to math. When it did, Huh tried not to give away how little he knew. "Somehow I was very good at pretending to understand what he was saying," Huh said. Indeed, Hironaka doesn't remember ever being aware of his would-be pupil's lack of formal training. "It's not anything I have a strong memory of. He was quite impressive to me," he said.

As the lunchtime conversations continued, their relationship grew. Huh graduated, and Hironaka stayed on at Seoul National for two more years. During that period, Huh began working on a master's degree in mathematics, mainly under Hironaka's direction. The two were almost always together. Hironaka would make occasional trips back home to Japan and Huh would go with him, carrying his bag through airports and even staying with Hironaka and his wife in their Kyoto apartment.

"I asked him if he wanted a hotel and he said he's not a hotel man. That's what he said. So he stayed in one corner of my apartment," Hironaka said.

In Kyoto and Seoul, Hironaka and Huh would go out to eat or take long walks, during which Hironaka would stop to photograph flowers. They became friends. "I liked him and he liked me, so we had that kind of non-mathematical chatting," Hironaka said.

Meanwhile, Hironaka continued to tutor Huh, working from concrete examples that Huh could understand rather than introducing him directly to general theories that might have been more than Huh could grasp. In particular, Hironaka taught Huh the nuances of singularity theory, the field where Hironaka had achieved his most famous results. Hironaka had also been trying for decades to find a proof of a major open problem—what's called the resolution of singularities in characteristic p. "It was a lifetime project for him, and that was principally what we talked about," Huh said. "Apparently he wanted me to continue this work."

In 2009, at Hironaka's urging, Huh applied to a dozen or so graduate schools in the U.S. His qualifications were slight: He hadn't majored in math; he'd taken few graduate-level classes; and his performance in those classes had been unspectacular. His case for admission rested largely on a recommendation from Hironaka. Most admissions committees were unimpressed. Huh got rejected at every school but one, the University of Illinois, Urbana-Champaign, where he enrolled in the fall of 2009.

A CRACK IN A GRAPH

At Illinois, Huh began the work that would ultimately lead him to a proof of Rota's conjecture. That problem was posed 47 years ago by the Italian mathematician Gian-Carlo Rota, and it deals with combinatorial objects—Tinkertoy-like constructions, like graphs, which are "combinations" of points and line segments glued together.

Consider a simple graph: a triangle.

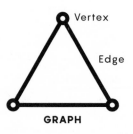

FIGURE 3.1

Mathematicians are interested in the following: How many different ways can you color the vertices of the triangle, given some number of colors and adhering to the rule that whenever two vertices are connected by an edge, they can't be the same color. Let's say you have q colors. Your options are as follows:

- q options for the first vertex, because when you're starting out you can use any color.

- $q-1$ options for the adjacent vertex, because you can use any color save the color you used to color the first vertex.

- $q-2$ options for the third vertex, because you can use any color save the two colors you used to color the first two vertices.

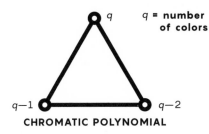

CHROMATIC POLYNOMIAL

FIGURE 3.2

The total number of colorings will be all options multiplied together, or in this case $q \times (q-1) \times (q-2) = q^3 - 3q^2 + 2q$.

That equation is called the chromatic polynomial for this graph, and it has some interesting properties.

Take the coefficients of each term: 1, −3 and 2. The absolute value of this sequence—1, 3, 2—has two properties in particular. The first is that it's "unimodal," meaning it only peaks once, and before that peak the sequence only ever rises, and after that peak it only ever falls.

The second property is that the sequence of coefficients is "log concave," meaning that any three consecutive numbers in the sequence follow this rule: The product of the outside two numbers is less than the square of the middle number. The sequence (1, 3, 5) satisfies this requirement ($1 \times 5 = 5$, which is smaller than 3^2), but the sequence (2, 3, 5) does not ($2 \times 5 = 10$, which is greater than 3^2).

You can imagine an infinite number of graphs—graphs with more vertices and more edges connected in any number of ways. Every one of these graphs has a unique chromatic polynomial. And in every graph that mathematicians have ever studied, the coefficients of its chromatic polynomial

THE CHROMATIC POLYNOMIAL OF A TRIANGLE

$$q \times (q - 1) \times (q - 2) = 1q^3 - 3q^2 + 2q$$

Coefficients of the formula

↓

1,3,2

The absolute value of this sequence

PROPERTY 1: SEQUENCE IS UNIMODAL

Sequence only peaks once, and before that peak the sequence
only ever rises, and after that peak it only ever falls.

1,**3**,2
↑
Peak

Examples of other
unimodal sequences:

1,2,3,4,**5**,4,3,2,1

2,3,5,7,**9**,8,7,6,5

PROPERTY 2: SEQUENCE IS LOG CONCAVE

Any three consecutive numbers in the sequence follow this rule:
The product of the outside two numbers is less
than the square of the middle number.

1,3,2 is log concave
(1 x 2 = 2 < 3²)

FIGURE 3.3

have always been both unimodal and log concave. That this fact always
holds is called "Read's conjecture." Huh would go on to prove it.

Read's conjecture is, in a sense, deeply counterintuitive. To understand
why, it helps to understand more about how graphs can be taken apart
and put back together. Consider a slightly more complicated graph—the
rectangle in figure 3.4.

The chromatic polynomial of the rectangle is harder to calculate than that of the triangle, but any graph can be broken up into subgraphs, which are easier to work with. Subgraphs are all the graphs you can make by deleting an edge (or edges) from the original graph, as in figure 3.5, or by contracting two vertices into one, as in figure 3.6.

The chromatic polynomial of the rectangle is equal to the chromatic polynomial of the rectangle with one edge deleted minus the chromatic polynomial of the triangle. This makes intuitive sense when you recognize that there should be more ways to color the rectangle with the deleted edge than the rectangle itself: The fact that the top two points aren't connected by an edge gives you more coloring flexibility (you can, for instance, color them

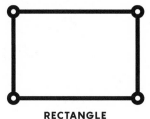

RECTANGLE

FIGURE 3.4

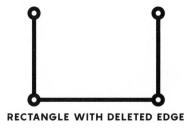

RECTANGLE WITH DELETED EDGE

FIGURE 3.5

RECTANGLE WITH CONTRACTED EDGE

FIGURE 3.6

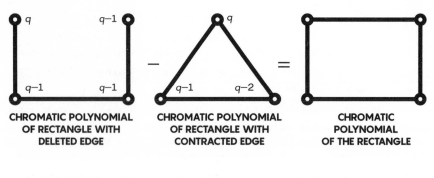

$$(q^4 - 3q^3 + 3q^2 - q) \; - \; (q^3 - 3q^2 + 2q) \; = \; q^4 - 4q^3 + 6q^2 - 3q$$

FIGURE 3.7

the same color, which you're not allowed to do when they're connected). Just how much flexibility does it give you? Precisely the number of coloring options for the triangle.

The chromatic polynomial for any graph can be defined in terms of the chromatic polynomials of subgraphs. And the coefficients of all of these chromatic polynomials are always log concave.

Yet when you add or subtract two log concave sequences, the resulting sequence is usually not itself log concave. Because of this, you'd expect log concavity to disappear in the process of combining chromatic polynomials. Yet it doesn't. Something else is going on. "This is what made people curious of this log concavity phenomenon," Huh said.

A SEARCH FOR HIDDEN STRUCTURE

Huh didn't know any of this when he arrived at Illinois. Most first-year graduate students spend more time in class than on their own research, but following his three-year apprenticeship with Hironaka, Huh had ideas that he wanted to pursue.

Through his first Midwestern winter, Huh developed techniques for applying singularity theory, the focus of his study with Hironaka, to graphs. In doing so, Huh found that when he constructed a singularity from a graph, he was suddenly able to use singularity theory to justify properties of the original graph—to explain, for instance, why the coefficients of a polynomial based on the graph would follow a log concave pattern.

This was interesting to Huh, so he searched the graph theory literature to see if others had previously explained these log concave patterns he was seeing. He discovered that to graph theorists, the patterns were still entirely mysterious.

"I noticed this pattern I'd observed was in fact a well-known conjecture in graph theory, Read's conjecture. In a sense I solved this problem without knowing the problem," Huh said.

Huh's inadvertent proof of Read's conjecture, and the way he combined singularity theory with graphs, could be seen as a product of his naïve approach to mathematics. He learned the subject mainly on his own and through informal study with Hironaka. People who have observed his rise over the last few years imagine that this experience left him less beholden to conventional wisdom about what kinds of mathematical approaches are worth trying. "If you look at mathematics as a kind of continent divided into countries, I think in June's case nobody really told him there were all these borders. He's definitely not constrained by any demarcations," said Robbert Dijkgraaf, the director of IAS.

Soon after he posted his proof of Read's conjecture, the University of Michigan invited Huh to give a talk on his result. On December 3, 2010, he addressed a room full of many of the same mathematicians who had rejected his graduate school application a year earlier. By this point Huh's talent was becoming evident to other mathematicians. Jesse Kass was a postdoctoral fellow in mathematics at Michigan at the time. Just before Huh's visit, a senior faculty member encouraged Kass to watch the talk because "30 years from now you can tell your grandchildren you saw Huh speak before he got famous," recalled Kass, who's now a professor at the University of South Carolina.

Huh's lecture did not disappoint.

"The talk was somehow very polished and very clear; it just went to the right points. It's a bit unusual for a beginning graduate student to give such clean talks," said Mircea Mustaţă, a mathematician at Michigan.

Following his talk, the Michigan faculty invited Huh to transfer, which he did in 2011. By that point he'd learned that Read's conjecture was a special case of a larger and more significant problem—Rota's conjecture.

Rota's conjecture is very similar to Read's conjecture, but instead of graphs it addresses more abstract combinatorial objects called "matroids" (a graph can be viewed as a particularly concrete type of matroid) and a different kind of equation that arises from each matroid, called the "characteristic polynomial." But the underlying point is the same: Rota's conjecture predicts that the coefficients of the characteristic polynomial for any matroid are always log concave.

The statement is simple, and evidence for it is abundant, but proving it—explaining why this log concavity occurs—is difficult. There's nothing about the matroids themselves that suggests why the log concavity would hold uniformly when you add or subtract the characteristic polynomials of submatroids (just as there's no obvious reason the log concavity would hold when you add or subtract the chromatic polynomials of graphs). Whenever you observe a pattern with no obvious cause, it's natural to start digging below the surface—to look for the roots that explain the tree. That's what Huh did when he and his collaborators began to attack Rota's conjecture.

"Log concavity is easy to observe in concrete examples," Huh said. "You just compute the sequence you're interested in and observe [that] it's there. But for some reason it's hard to justify why this is happening."

Initially Huh looked for ways to extend the techniques from singularity theory that he had used for Read's conjecture, but he quickly found that they didn't work in the more abstract realm of matroids.

This failure left him looking for some other kind of structure, hidden beneath the surface of matroids, that could explain the way they behaved mathematically.

CROSSING BOUNDARIES

Some of the biggest leaps in understanding occur when someone extends a well-established theory in one area to seemingly unrelated phenomena in another. Think, for example, about gravitation. People have always understood that objects fall to the ground when released from a height; the heavens became far more intelligible when Newton realized the same dynamic explained the motion of the planets.

In math the same kind of transplantation occurs all the time. In his widely cited 1994 essay "On Proof and Progress in Mathematics," the influential mathematician William Thurston explained that there are dozens of different ways to think of the concept of the "derivative."[2] One is the way you learn it in calculus—the derivative as a measure of infinitesimal change in a function. But the derivative appears in other guises: as the slope of a line tangent to the graph of a function, or as the instantaneous speed given by a function at a specific time. "This is a list of different ways of *thinking about* or *conceiving of* the derivative, rather than a list of different *logical definitions*," Thurston wrote.

Huh's work on Rota's conjecture involved a reconception of a venerable area of mathematics called Hodge theory. Hodge theory was developed in the 1930s by the Scottish mathematician William Hodge. To call it a "theory" is simply to say that it's the study of some particular thing, just as you could say that "right triangle theory" is the study of right triangles. In the case of Hodge theory, the objects of interest are called the "cohomology rings of smooth projective algebraic varieties."

It's hard to overstate how little Hodge theory would seem to relate to graphs or matroids. The cohomology rings in Hodge theory arise from smooth functions that come packaged with a concept of the infinite. By contrast, combinatorial objects like graphs and matroids are purely discrete objects—assemblages of dots and sticks. To ask what Hodge theory means in the context of matroids is a little like asking how to take the square root of a sphere. The question doesn't appear to make any sense.

Yet there was good reason to ask. In the more than 60 years since Hodge theory was proposed, mathematicians have found a number of instances of Hodge-type structures appearing in settings far removed from their original algebraic context. It's as if the Pythagorean relationship, once thought to be the exclusive provenance of right triangles, also turned out to describe the distribution of prime numbers.

"There is some feeling that these structures, whenever they exist, are fundamental. They explain facts about your mathematical structure that are hard to explain by any other means," Huh said.

Some of these new settings have been combinatorial, which encouraged Huh to wonder whether relationships from Hodge theory might underlie these log concave patterns. Searching for a familiar math concept in a foreign land is not easy, though. In fact, it's a bit like searching for extraterrestrial life—you might have ideas about signature characteristics of life, hints you might use to guide your hunt, but it's still hard to anticipate what a new life-form might look like.

A PARTNERSHIP GROWS

In recent years Huh has done much of his most important work with two collaborators—Eric Katz, a mathematician at Ohio State University, and Karim Adiprasito, a mathematician at the Hebrew University of Jerusalem. They make an unusual trio.

Adiprasito initially wanted to be a chef and spent time backpacking around India before settling down in combinatorics, the area of mathematics that's home to graph theory and problems like Rota's conjecture. He liked math in high school but turned away from it because "I felt it wasn't creative enough for me," he said. Katz has a frenetic mind and obsessively detailed knowledge of indie rock bands, developed in his earlier years as a college radio DJ. Of the three collaborators, he is the closest to having a typical math pedigree, and he views himself as a kind of interpreter between the creative ideas of the would-be poet and the would-be chef.

"Karim has these amazing ideas that come out of nowhere, and June sort of has this beautiful vision of how math should go," Katz said. "It's often hard to incorporate Karim's ideas into June's vision, and maybe some of what I do is talk to Karim and translate his ideas into something closer to math."

Katz became aware of Huh's work in 2011, after Huh proved Read's conjecture but before he'd made any progress proving Rota's conjecture. Katz read Huh's proof of Read's conjecture and observed that if he cut out one particular step in the argument, he could apply the methods from that paper to get a partial proof of Rota's conjecture. He contacted Huh, and over the course of just a few months the two wrote a paper (published in 2012) that explained log concavity for a small class of matroids called "realizable" matroids.

Yet that paper didn't solve the hardest part of Rota's conjecture—proving log concavity for "nonrealizable" matroids, which comprise the vast majority of all matroids. Hodge theory, remember, was defined originally in the 1950s on objects called the "cohomology rings of algebraic varieties." If you want to prove that Hodge-type structures explain phenomena observed in matroids, you need to find a way of explaining how something like a cohomology ring can be distilled out of a matroid. With realizable matroids, there was a very straightforward way to do this, which is why Huh and Katz's proof for that piece of Rota's conjecture came so quickly. But with nonrealizable matroids, there was no obvious way to instantiate a cohomology ring—they were like a language without a word for that concept.

For four years Huh and Katz tried and failed to find a way to define what a Hodge structure would mean in the context of nonrealizable matroids. During that time they determined that one particular aspect of Hodge theory—known as the Hodge index theorem—would be enough by itself to explain log concavity, but there was a catch: They couldn't find a way to actually prove that the Hodge index theorem was true for matroids.

That's when Adiprasito entered the picture. In 2015 he traveled to IAS and visited Huh. Adiprasito realized that while the Hodge index theorem alone would explain log concavity, the way to prove the Hodge index theorem for matroids was to try and prove a larger set of ideas from Hodge theory that includes the Hodge index theorem—what the three collaborators refer to as the "Kähler package."

"I told June and Eric there is a way to actually prove this in a purely combinatorial setting," Adiprasito said. "Then it actually was quick that we came up with a plan. I think they asked the question and I provided the technique."

This technique provided a full proof of Rota's conjecture. The trio posted their work online in November 2015, and since then it has rippled through the math world.[3] Their work provides a fully combinatorial vision of Hodge theory, which in turn provides a whole new way to approach open problems in combinatorics.

The work has also elevated Huh's profile. In addition to his new position at IAS, Huh is frequently mentioned as a strong contender for the Fields Medal, which is given every four years to the most accomplished mathematicians under the age of 40.

THE STUDENT SETS OFF ON HIS OWN

Back in 2012, Huh went to Seoul National University to give a talk on his recent proof of Read's conjecture. Hironaka was in the audience, and he recalls being surprised to learn that singularity theory had applications to graphs. Afterward, he asked Huh if this new work marked a change in his research interests.

"I remember I asked him if he's completely in graph theory kinds of things and has lost interest in singularities. He said no, he's still interested in singularities," Hironaka said.

Huh remembers that conversation, too. It took place at a time when he was indeed setting out in a whole new direction in mathematics. He thinks maybe he just wasn't ready to say that out loud—especially to the man who changed his life. "That was the point that I was going off the road," he said. "I think he sensed that and still, I am off the road. Maybe there was some psychological force that made me not want to admit I'd completely left singularity theory behind."

Huh and Hironaka have not seen each other since. Hironaka is now 87. He's retired but continues to work toward a proof of the problem in singularity theory that has occupied him for decades. In March 2017 he posted a long paper to his old faculty webpage at Harvard University that he says provides a proof. Other mathematicians, including Huh, have taken a preliminary look at the work but have not yet verified that the proof holds. It's hard for Hironaka to travel, but he wishes it were easier for him to see Huh again. "I only hear about him," Hironaka said.

Over coffee one afternoon at Huh's apartment on the IAS campus, I asked him how he feels about not pursuing the research track Hironaka may have hoped for him. He thought for a moment, then said he feels guilty.

"A lot of the time with Hironaka I sort of had to fake my understanding," he said. "This lack of mathematical background has prevented me from going on to serious research with him. This has left a sort of long-term homework in me."

At the same time, Huh regards the distance he has traveled from his mathematical roots as a good and maybe necessary step in the development of his work. As we parted on a street corner in Princeton, he remarked, "I need space to think," before heading into the quiet confines of IAS. He found his own way into mathematics, and now that he's there, he's going to find his own path through it.

A LONG-SOUGHT PROOF, FOUND AND ALMOST LOST

Natalie Wolchover

A s he was brushing his teeth on the morning of July 17, 2014, Thomas Royen, a little-known retired German statistician, suddenly lit upon the proof of a famous conjecture at the intersection of geometry, probability theory and statistics that had eluded top experts for decades.

Known as the Gaussian correlation inequality (GCI), the conjecture originated in the 1950s, was posed in its most elegant form in 1972 and has held mathematicians in its thrall ever since. "I know of people who worked on it for 40 years," said Donald Richards, a statistician at Pennsylvania State University. "I myself worked on it for 30 years."

Royen hadn't given the Gaussian correlation inequality much thought before the "raw idea" for how to prove it came to him over the bathroom sink. Formerly an employee of a pharmaceutical company, he had moved on to a small technical university in Bingen, Germany, in 1985 in order to have more time to improve the statistical formulas that he and other industry statisticians used to make sense of drug-trial data. In July 2014, still at work on his formulas as a 67-year-old retiree, Royen found that the GCI could be extended into a statement about statistical distributions he had long specialized in. On the morning of the 17th, he saw how to calculate a key derivative for this extended GCI that unlocked the proof. "The evening of this day, my first draft of the proof was written," he said.

Not knowing LaTeX, the word processer of choice in mathematics, he typed up his calculations in Microsoft Word, and the following month he posted his paper to the academic preprint site arxiv.org.[1] He also sent it to Richards, who had briefly circulated his own failed attempt at a proof of the GCI a year and a half earlier. "I got this article by email from him," Richards said. "And when I looked at it I knew instantly that it was solved."

Upon seeing the proof, "I really kicked myself," Richards said. Over the decades, he and other experts had been attacking the GCI with increasingly sophisticated mathematical methods, certain that bold new ideas in convex geometry, probability theory or analysis would be needed to prove it. Some mathematicians, after years of toiling in vain, had come to suspect the inequality was actually false. In the end, though, Royen's proof was short and simple, filling just a few pages and using only classic techniques. Richards was shocked that he and everyone else had missed it. "But on the other hand I have to also tell you that when I saw it, it was with relief," he said. "I remember thinking to myself that I was glad to have seen it before I died." He laughed. "Really, I was so glad I saw it."

Richards notified a few colleagues and even helped Royen retype his paper in LaTeX to make it appear more professional. But other experts whom Richards and Royen contacted seemed dismissive of his dramatic claim. False proofs of the GCI had been floated repeatedly over the decades, including two that had appeared on arxiv.org since 2010. Bo'az Klartag of the Weizmann Institute of Science and Tel Aviv University recalls receiving the batch of three purported proofs, including Royen's, in an email from a colleague in 2015. When he checked one of them and found a mistake, he set the others aside for lack of time. For this reason and others, Royen's achievement went unrecognized.

Proofs of obscure provenance are sometimes overlooked at first, but usually not for long: A major paper like Royen's would normally get submitted and published somewhere like the *Annals of Statistics*, experts said, and then everybody would hear about it. But Royen, not having a career to advance, chose to skip the slow and often demanding peer-review process typical of top journals. He opted instead for quick publication in the *Far East Journal of Theoretical Statistics*, a periodical based in Allahabad, India, that was largely unknown to experts and which, on its website, rather suspiciously listed Royen as an editor. (He had agreed to join the editorial board the year before.)

With this red flag emblazoned on it, the proof continued to be ignored. Finally, in December 2015, the Polish mathematician Rafał Latała and his student Dariusz Matlak put out a paper advertising Royen's proof, reorganizing it in a way some people found easier to follow.[2] Word still took its sweet time to get around. Tilmann Gneiting, a statistician at the Heidelberg Institute for Theoretical Studies, just 65 miles from Bingen, said he was shocked to learn in July 2016, two years after the fact, that the GCI had been proved. The statistician Alan Izenman, of Temple University in Philadelphia, still hadn't heard about the proof when asked for comment in early 2017.

No one is quite sure how, in the 21st century, news of Royen's proof managed to travel so slowly. "It was clearly a lack of communication in an age where it's very easy to communicate," Klartag said.

"But anyway, at least we found it," he added—and "it's beautiful."

In its most famous form, formulated in 1972, the GCI links probability and geometry: It places a lower bound on a player's odds in a game of darts, including hypothetical dart games in higher dimensions.[3]

Imagine two convex polygons, such as a rectangle and a circle, centered on a point that serves as the target. Darts thrown at the target will land in a bell curve or "Gaussian distribution" of positions around the center point. The Gaussian correlation inequality says that the probability that a dart will land inside both the rectangle and the circle is always as high as or higher than the individual probability of its landing inside the rectangle multiplied by the individual probability of its landing in the circle. In plainer terms, because the two shapes overlap, striking one increases your chances of also striking the other. The same inequality was thought to hold for any two convex symmetrical shapes with any number of dimensions centered on a point.

Special cases of the GCI have been proved—in 1977, for instance, Loren Pitt of the University of Virginia established it as true for two-dimensional convex shapes—but the general case eluded all mathematicians who tried to prove it.[4] Pitt had been trying since 1973, when he first heard about the inequality over lunch with colleagues at a meeting in Albuquerque, New Mexico. "Being an arrogant young mathematician...I was shocked that grown men who were putting themselves off as respectable math and science people didn't know the answer to this," he said. He locked himself in his motel room and was sure he would prove or disprove the conjecture before coming out. "Fifty years or so later I still didn't know the answer," he said.

Despite hundreds of pages of calculations leading nowhere, Pitt and other mathematicians felt certain—and took his 2-D proof as evidence—that the convex geometry framing of the GCI would lead to the general proof. "I had developed a conceptual way of thinking about this that perhaps I was overly wedded to," Pitt said. "And what Royen did was kind of diametrically opposed to what I had in mind."

Royen's proof harkened back to his roots in the pharmaceutical industry, and to the obscure origin of the Gaussian correlation inequality itself. Before it was a statement about convex symmetrical shapes, the GCI was conjectured in 1959 by the American statistician Olive Dunn as a formula for calculating "simultaneous confidence intervals," or ranges that multiple variables are all estimated to fall in.[5]

Suppose you want to estimate the weight and height ranges that 95 percent of a given population fall in, based on a sample of measurements. If you plot people's weights and heights on an x–y plot, the weights will form a Gaussian bell-curve distribution along the x-axis, and heights will form a bell curve along the y-axis. Together, the weights and heights follow a two-dimensional bell curve. You can then ask, what are the weight and height ranges—call them $-w < x < w$ and $-h < y < h$—such that 95 percent of the population will fall inside the rectangle formed by these ranges?

If weight and height were independent, you could just calculate the individual odds of a given weight falling inside $-w < x < w$ and a given height falling inside $-h < y < h$, then multiply them to get the odds that both conditions are satisfied. But weight and height are correlated. As with darts and overlapping shapes, if someone's weight lands in the normal range, that person is more likely to have a normal height. Dunn, generalizing an inequality posed three years earlier, conjectured the following: The probability that both Gaussian random variables will simultaneously fall inside the rectangular region is always greater than or equal to the product of the individual probabilities of each variable falling in its own specified range. (This can be generalized to any number of variables.) If the variables are independent, then the joint probability equals the product of the individual probabilities. But any correlation between the variables causes the joint probability to increase.

Royen found that he could generalize the GCI to apply not just to Gaussian distributions of random variables but to more general statistical spreads related to the squares of Gaussian distributions, called gamma distributions, which are used in certain statistical tests. "In mathematics, it occurs frequently that a seemingly difficult special problem can be solved by answering a more general question," he said.

Royen represented the amount of correlation between variables in his generalized GCI by a factor we might call C, and he defined a new function whose value depends on C. When $C = 0$ (corresponding to independent variables like weight and eye color), the function equals the product of the separate probabilities. When you crank up the correlation to the maximum, $C = 1$, the function equals the joint probability. To prove that the latter is bigger than the former and the GCI is true, Royen needed to show that his function always increases as C increases. And it does so if its derivative, or rate of change, with respect to C is always positive.

His familiarity with gamma distributions sparked his bathroom-sink epiphany. He knew he could apply a classic trick to transform his function into a simpler function. Suddenly, he recognized that the derivative of this transformed function was equivalent to the transform of the derivative of the original function. He could easily show that the latter derivative was always positive, proving the GCI. "He had formulas that enabled him to pull off his magic," Pitt said. "And I didn't have the formulas."

Any graduate student in statistics could follow the arguments, experts said. Royen said he hopes the "surprisingly simple proof … might encourage young students to use their own creativity to find new mathematical theorems," since "a very high theoretical level is not always required."

Some researchers, however, still want a geometric proof of the GCI, which would help explain strange new facts in convex geometry that are only de facto implied by Royen's analytic proof. In particular, Pitt said, the GCI defines an interesting relationship between vectors on the surfaces of overlapping convex shapes, which could blossom into a new subdomain of convex geometry. "At least now we know it's true," he said of the vector relationship. But "if someone could see their way through this geometry we'd understand a class of problems in a way that we just don't today."

Beyond the GCI's geometric implications, Richards said a variation on the inequality could help statisticians better predict the ranges in which variables like stock prices fluctuate over time. In probability theory, the GCI proof now permits exact calculations of rates that arise in "small-ball" probabilities, which are related to the random paths of particles moving in a fluid. Richards says he has conjectured a few inequalities that extend the GCI, and which he might now try to prove using Royen's approach.

Royen's main interest is in improving the practical computation of the formulas used in many statistical tests—for instance, for determining whether a drug causes fatigue based on measurements of several variables, such as patients' reaction time and body sway. He said that his extended GCI does indeed sharpen these tools of his old trade, and that some of his other recent work related to the GCI has offered further improvements. As for the proof's muted reception, Royen wasn't particularly disappointed or surprised. "I am used to being frequently ignored by scientists from [top-tier] German universities," he wrote in an email. "I am not so talented for 'networking' and many contacts. I do not need these things for the quality of my life."

The "feeling of deep joy and gratitude" that comes from finding an important proof has been reward enough. "It is like a kind of grace," he said. "We can work for a long time on a problem and suddenly an angel—[which] stands here poetically for the mysteries of our neurons—brings a good idea."

"OUTSIDERS" CRACK 50-YEAR-OLD MATH PROBLEM

Erica Klarreich

In 2008, Daniel Spielman told his Yale University colleague Gil Kalai about a computer science problem he was working on, concerning how to "sparsify" a network so that it has fewer connections between nodes but still preserves the essential features of the original network.

Network sparsification has applications in data compression and efficient computation, but Spielman's particular problem suggested something different to Kalai. It seemed connected to the famous Kadison–Singer problem, a question about the foundations of quantum physics that had remained unsolved for almost 50 years.

Over the decades, the Kadison–Singer problem had wormed its way into a dozen distant areas of mathematics and engineering, but no one seemed to be able to crack it. The question "defied the best efforts of some of the most talented mathematicians of the last 50 years," wrote Peter Casazza and Janet Tremain of the University of Missouri in Columbia, in a 2014 survey article.[1]

As a computer scientist, Spielman knew little of quantum mechanics or the Kadison–Singer problem's allied mathematical field, called C*-algebras. But when Kalai, whose main institution is the Hebrew University of Jerusalem, described one of the problem's many equivalent formulations, Spielman realized that he himself might be in the perfect position to solve it. "It seemed so natural, so central to the kinds of things I think about," he said. "I thought, 'I've got to be able to prove that.'" He guessed that the problem might take him a few weeks.

Instead, it took him five years. In 2013, working with his postdoc Adam Marcus, now at Princeton University, and his graduate student Nikhil Srivastava, now at the University of California, Berkeley, Spielman finally succeeded.[2] Word spread quickly through the mathematics community that one of the paramount problems in C*-algebras and a host of other fields had been solved by three outsiders—computer scientists who had barely a nodding acquaintance with the disciplines at the heart of the problem.

Mathematicians in these disciplines greeted the news with a combination of delight and hand-wringing. The solution, which Casazza and Tremain called "a major achievement of our time," defied expectations about how the problem would be solved and seemed bafflingly foreign. Afterward, the experts in the Kadison–Singer problem worked hard to assimilate the ideas of the proof. Spielman, Marcus and Srivastava "brought a bunch of tools into this problem that none of us had ever heard of," Casazza said. "A lot of us loved this problem and were dying to see it solved, and we had a lot of trouble understanding how they solved it."

"The people who have the deep intuition about why these methods work are not the people who have been working on these problems for a long time," said Terence Tao of UCLA, who has been following these developments. Mathematicians have held several workshops to unite these disparate camps, but the proof may take several more years to digest, Tao said. "We don't have the manual for this magic tool yet."

Computer scientists, however, have been quick to exploit the new techniques. In 2014, for instance, two researchers parlayed these tools into a major leap forward in understanding the famously difficult traveling salesman problem. There are certain to be more such advances, said Assaf Naor, a mathematician at Princeton who works in areas related to the Kadison–Singer problem. "This is too profound to not have many more applications."

A COMMON PROBLEM

The question Richard Kadison and Isadore Singer posed in 1959 asks how much it is possible to learn about a "state" of a quantum system if you have complete information about that state in a special subsystem. Inspired by an informally worded comment by the legendary physicist Paul Dirac, their question builds on Werner Heisenberg's uncertainty principle, which says that certain pairs of attributes, like the position and the momentum of a particle, cannot simultaneously be measured to arbitrary precision.

Kadison and Singer wondered about subsystems that contain as many different attributes (or "observables") as can compatibly be measured at the same time. If you have complete knowledge of the state of such a subsystem, they asked, can you deduce the state of the entire system?

In the case where the system you're measuring is a particle that can move along a continuous line, Kadison and Singer showed that the answer is no: There can be many different quantum states that all look the same from the point of view of the observables you can simultaneously measure.[3] "It is as if many different particles have exactly the same location simultaneously—in a sense, they are in parallel universes," Kadison wrote by email, although he cautioned that it's not yet clear whether such states can be realized physically.

Kadison and Singer's result didn't say what would happen if the space in which the particle lives is not a continuous line, but is instead some choppier version of the line—if space is "granular," as Kadison put it. This is the question that came to be known as the Kadison–Singer problem.

Based on their work in the continuous setting, Kadison and Singer guessed that in this new setting the answer would again be that there are parallel universes. But they didn't go so far as to state their guess as a conjecture—a wise move, in hindsight, since their gut instinct turned out to be wrong. "I'm happy I've been careful," Kadison said.

Kadison and Singer—now at the University of Pennsylvania and the Massachusetts Institute of Technology (emeritus), respectively—posed their question at a moment when interest in the philosophical foundations of quantum mechanics was entering a renaissance. Although some physicists were promoting a "shut up and calculate" approach to the discipline, other, more mathematically inclined physicists pounced on the Kadison–Singer problem, which they understood as a question about C*-algebras, abstract structures that capture the algebraic properties not just of quantum systems but also of the random variables used in probability theory, the blocks of numbers called matrices and regular numbers.

C*-algebras are an esoteric subject—"the most abstract nonsense that exists in mathematics," in Casazza's words. "Nobody outside the area knows much about it." For the first two decades of the Kadison–Singer problem's existence, it remained ensconced in this impenetrable realm.

Then in 1979, Joel Anderson, now an emeritus professor at Pennsylvania State University, popularized the problem by proving that it is equivalent to an easily stated question about when matrices can be broken down into simpler chunks.[4] Matrices are the core objects in linear algebra, which is used to study mathematical phenomena whose behavior can be captured by lines, planes and higher-dimensional spaces. So suddenly, the Kadison–Singer problem was everywhere. Over the decades that followed, it emerged as the key problem in one field after another.

Because there tended to be scant interaction between these disparate fields, no one realized just how ubiquitous the Kadison–Singer problem had become until Casazza found that it was equivalent to the most important problem in his own area of signal processing. The problem concerned whether the processing of a signal can be broken down into smaller, simpler parts. Casazza dived into the Kadison–Singer problem, and in 2005, he, Tremain and two co-authors wrote a paper demonstrating that it was equivalent to the biggest unsolved problems in a dozen areas of math and engineering.[5] A solution to any one of these problems, the authors showed, would solve them all.

One of the many equivalent formulations they wrote about had been devised just a few years earlier by Nik Weaver, of Washington University in St. Louis.[6] Weaver's version distilled the problem down to a natural-sounding question about when it is possible to divide a collection of vectors into two groups that each point in roughly the same set of directions as the original collection. "It's a beautiful problem that brought out the core combinatorial problem" at the heart of the Kadison–Singer question, Weaver said.

So Weaver was surprised when—apart from the mention in Casazza's survey and one other paper that expressed skepticism about his approach—his formulation seemed to meet with radio silence. He thought no one had noticed his paper, but in fact it had attracted the attention of just the right people to solve it.

ELECTRICAL PROPERTIES

When Spielman learned about Weaver's conjecture in 2008, he knew it was his kind of problem. There's a natural way to switch between networks and collections of vectors, and Spielman had spent the preceding several years building up a powerful new approach to networks by viewing them as physical objects. If a network is thought of as an electrical circuit, for example, then the amount of current that runs through a given edge (instead of finding alternate routes) provides a natural way to measure that edge's importance in the network.

Spielman discovered Weaver's conjecture after Kalai introduced him to another form of the Kadison–Singer problem, and he realized that it was nearly identical to a simple question about networks: When is it possible to divide up the edges of a network into two categories—say, red edges and blue edges—so that the resulting red and blue networks have similar electrical properties to the whole network?

It's not always possible to do this. For instance, if the original network consists of two highly connected clusters that are linked to each other by a single edge, then that edge has an outsize importance in the network. So if that critical edge is colored red, then the blue network can't have similar electrical properties to the whole network. In fact, the blue network won't even be connected.

Weaver's problem asks whether this is the only type of obstacle to breaking down networks into similar but smaller ones. In other words, if there are enough ways to get around in a network—if no individual edge is too important—can the network be broken down into two subnetworks with similar electrical properties?

Spielman, Marcus and Srivastava suspected that the answer was yes, and their intuition did not just stem from their previous work on network sparsification. They also ran millions of simulations without finding any counterexamples. "A lot of our stuff was led by experimentation," Marcus said. "Twenty years ago, the three of us sitting in the same room would not have solved this problem."

The simulations convinced them that they were on the right track, even as the problem raised one stumbling block after another. And they kept making spurts of progress, enough to keep them hooked. When Marcus' postdoctoral fellowship expired at the end of the team's fourth year working on the problem, he elected to leave academia temporarily and join a local startup called Crisply rather than leave New Haven. "I worked for my company four days a week, and then once a week or so I would go to Yale," he said.

A network's electrical properties are governed by a special equation called the network's "characteristic polynomial." As the trio performed computer experiments on these polynomials, they found that the equations seemed to have hidden structure: Their solutions were always real numbers (as opposed to complex numbers), and, surprisingly, adding these polynomials together always seemed to result in a new polynomial with that same property. "These polynomials were doing more than we gave them credit for," Marcus said. "We used them as a way of transferring knowledge, but really the polynomials seemed to be containing knowledge themselves."

Piece by piece, the researchers developed a new technique for working with so-called "interlacing polynomials" to capture this underlying structure, and finally, on June 17, 2013, Marcus sent an email to Weaver, who had been his undergraduate advisor at Washington University 10 years earlier. "I hope you remember me," Marcus wrote. "The reason I am writing is because we…think we have solved your conjecture (the one that you showed was equivalent to Kadison–Singer)." Within days, news of the team's achievement had spread across the blogosphere.

The proof, which has since been thoroughly vetted, is highly original, Naor said. "What I love about it is just this feeling of freshness," he said.

"That's why we want to solve open problems—for the rare events when somebody comes up with a solution that's so different from what was before that it just completely changes our perspective."

Computer scientists have already applied this new point of view to the "asymmetric" traveling salesman problem. In the traveling salesman problem, a salesman must travel through a series of cities, with the goal of minimizing the total distance traveled; the asymmetric version includes situations in which the distance from A to B differs from the distance from B to A (for instance, if the route includes one-way streets).

The best-known algorithm for finding approximate solutions to the asymmetric problem dates back to 1970, but no one knew how good its approximations were.[7] Now, using ideas from the proof of the Kadison–Singer problem, Nima Anari, of the University of California, Berkeley, and Shayan Oveis Gharan, of the University of Washington in Seattle, have shown that this algorithm performs exponentially better than people had realized.[8] The new result is "major, major progress," Naor said.

The proof of the Kadison–Singer problem implies that all the constructions in its dozen incarnations can, in principle, be carried out—quantum knowledge can be extended to full quantum systems, networks can be decomposed into electrically similar ones, matrices can be broken into simpler chunks. The proof won't change what quantum physicists do, but it could have applications in signal processing, since it implies that collections of vectors used to digitize signals can be broken down into smaller frames that can be processed faster. The theorem "has potential to affect some important engineering problems," Casazza said.

But there's a big gulf between principle and practice. The proof establishes that these various constructions exist, but it doesn't say how to carry them out. At present, Casazza said, "there isn't a chance in hell" of pulling a useful algorithm out of the proof. However, now that mathematicians know that the question has a positive answer, he hopes that a constructive proof will be forthcoming—not to mention a proof that mathematicians in his field can actually understand. "All of us were completely convinced it had a negative answer, so none of us was actually trying to prove it," he said.

Mathematicians in the fields in which the Kadison–Singer problem has been prominent may feel wistful that three outsiders came in and solved "their" central problem, but that's not what really happened, Marcus said. "The only reason we could even try to solve such a problem is because people in that field had already removed all the hardness that was happening" in C*-algebras, he said. "There was just one piece left, and that piece wasn't a problem they had the techniques to solve. I think the reason why this problem lasted 50 years is because it really had two parts that were hard."

Throughout the five years he spent working on the Kadison–Singer problem, Marcus said, "I don't think I could have told you what the problem was in the C*-algebra language, because I had no clue." The fact that he, Srivastava and Spielman were able to solve it "says something about what I hope will be the future of mathematics," he said. When mathematicians import ideas across fields, "that's when I think these really interesting jumps in knowledge happen."

MATHEMATICIANS TAME ROGUE WAVES, LIGHTING UP FUTURE OF LEDS

Kevin Hartnett

I n the 1950s, Philip Anderson, a physicist at Bell Laboratories, discovered a strange phenomenon. In some situations where it seems as though waves should advance freely, they just stop—like a tsunami halting in the middle of the ocean.

Anderson won the 1977 Nobel Prize in Physics for his discovery of what is now called Anderson localization, a term that refers to waves that stay in some "local" region rather than propagating the way you'd expect. He studied the phenomenon in the context of electrons moving through impure materials (electrons behave as both particles and waves), but under certain circumstances it can happen with other types of waves as well.

Even after Anderson's discovery, much about localization remained mysterious. Although researchers were able to prove that localization does indeed occur, they had a very limited ability to predict when and where it might happen. It was as if you were standing on one side of a room, expecting a sound wave to reach your ear, but it never did. Even if, after Anderson, you knew that the reason it didn't was that it had localized somewhere on its way, you'd still like to figure out exactly where it had gone. And for decades, that's what mathematicians and physicists struggled to explain.

This is where Svitlana Mayboroda comes in. Mayboroda, 37, is a mathematician at the University of Minnesota. In 2012, she began to untangle the long-standing puzzle of localization. She came up with a mathematical formula called the "landscape function" that predicts exactly where waves will localize and what form they'll take when they do.

"You want to know how to find these areas of localization," Mayboroda said. "The naive approach is difficult. The landscape function magically gives a way of doing it."

Her work began in the realm of pure mathematics, but unlike most mathematical advances, which might find a practical use after decades, if ever, her work is already being applied by physicists. In particular, LED lights—or light-emitting diodes—depend on the phenomenon of localization. They light up when electrons in a semiconducting material, having started out in a position of higher energy, get trapped (or "localize") in a position of lower energy and emit the difference as a photon of light. LEDs are still a work in progress: Engineers need to build LEDs that more efficiently convert electrons into light, if the devices are to become the future of artificial lighting, as many expect they will. If physicists can gain a better understanding of the mathematics of localization, engineers can build better LEDs—and with the help of Mayboroda's mathematics, that effort is already under way.

ROGUE WAVES

Localization is not an intuitive concept. Imagine you stood on one side of a room and watched someone ring a bell, only the sound never reached your ears. Now imagine that the reason it didn't is that the sound had fallen into an architectural trap, like the sound of the sea bottled in a shell.

Of course, in an ordinary room that never happens: Sound waves propagate freely until they hit your eardrums, or get absorbed into the walls, or dissipate in collisions with molecules in the air. But Anderson realized that when waves move through highly complex or disordered spaces, like a room with very irregular walls, the waves can trap themselves in place.

Anderson studied localization in electrons moving through a material. He realized that if the material is well-ordered, like a crystal, with its atoms evenly distributed, the electrons move freely as waves. But if the material's atomic structure is more random—with some atoms here, and a whole bunch over there, as is the case in many industrially manufactured alloys—then the electron waves scatter and reflect in highly complicated ways that can lead the waves to disappear altogether.

"Disorder is inevitable in the way these materials are created, there's no way to escape it," said Marcel Filoche, a physicist at the École Polytechnique outside Paris and a close collaborator of Mayboroda's. "The only thing to hope is that you can play with it, control it."

Physicists have long understood that localization is related to wave interference. If the peaks of one wave align with the troughs of another, you get destructive interference, and the two waves will cancel each other out.

Localization takes place when waves cancel each other out everywhere except in a few isolated places. For such nearly complete cancellation to occur, you need the waves to be moving in a complicated space that breaks the waves into a huge variety of sizes. These waves then interfere with each other in a bewildering number of ways. And, just as you can combine every color to get black, when you combine such a complicated mix of sound waves you get silence.

The principle is simple. The calculations are not. Understanding localization has always required simulating the infinite variety of wave sizes and exploring every possible way those waves could interfere with each other. It's an overwhelming calculation that can take researchers months to carry out in the kinds of three-dimensional materials physicists actually want to understand. With some materials, it's altogether impossible.

Unless you have the landscape function.

THE LAY OF THE LANDSCAPE

In 2009 Mayboroda went to France and presented research she'd been doing on the mathematics of thin plates. She explained that when the plates have a complicated shape, and you apply some pressure from one side, the plates may flex in very irregular ways—bulging out in unexpected places, while remaining almost flat in others.

Filoche was in the audience. He had spent more than a decade studying the localization of vibrations, and his research had led to the construction of a prototype noise-abating barrier called "Fractal Wall" for use along highways. After Mayboroda's talk, the two started speculating whether the irregular bulging patterns in Mayboroda's plates might be related to the way Filoche's vibrations localized in some places and disappeared in others.

Over the next three years they found that the two phenomena were indeed related. In a 2012 paper, Filoche and Mayboroda introduced a way to mathematically perceive the terrain the way a wave would see it.[1] The resulting "landscape" function interprets information about the geometry and material a wave is moving through and uses it to draw the boundaries of localization. Previous efforts to pinpoint localized waves had failed due to the complexity of considering all possible waves, but Mayboroda and Filoche found a way to reduce the problem to a single mathematical expression.

To see how the landscape function works, think about a thin plate with a complex outer boundary. Imagine striking it with a rod. It might remain silent in some places and ring in others. How do you know what's going to happen and where?

The landscape function considers how the plate flexes under uniform pressure. The places it bulges when placed under pressure aren't visible, but the vibrations perceive those bulges and so does the landscape function:

The bulges are where the plate will ring, and the lines around the bulges are precisely the lines of localization drawn by the function.

"Imagine a plate, subject it to air pressure on one side, push it, then measure the nonuniformity of how much points are bulging out. That's the landscape function, that's it," said David Jerison, a mathematician at MIT and a collaborator on the landscape function work.

Following their 2012 paper, Mayboroda and Filoche looked for ways to extend the landscape function from mechanical vibrations to the quantum world of electron waves.

Electrons are unique among wavelike phenomena. Instead of picturing a wave, think of them as having more or less energy depending on where they're located in the atomic structure of a material. For a given material there's a map, called the potential (as in "potential energy"), that tells you the energy. The potential is relatively easy to draw for materials like conductors that have an orderly atomic structure, but it's very difficult to calculate in materials with highly irregular atomic structures. These disordered materials are precisely the ones in which electron waves will undergo localization.

"The randomness of the material makes prediction of the potential map very difficult," Filoche explained in an email. "Moreover, this potential map also depends on the location of the moving electrons, while the motion of the electrons depends in turn on the potential."

Another challenge in drawing the potential for a disordered material is that when waves localize in a region, they're not actually confined completely to that area, and they gradually vanish as they get farther away from the localization region. In mechanical systems, such as a vibrating plate, these distant traces of a wave can be safely ignored. But in quantum systems filled with hypersensitive electrons, those traces matter.

"If you have an electron here and another there and they're localized in different places, the only way they'll interact will be by their exponentially decaying tails. For interacting quantum systems, you absolutely need [to be able to describe] this," Filoche said.

Over the next five years Filoche and Mayboroda brought in additional collaborators and improved the landscape function's predictive power. Together with Jerison, Douglas Arnold of the University of Minnesota and Guy David of the University of Paris-South, they have developed a new version of the landscape function—which, in simple terms, is the reciprocal of the original one—that exactly predicts where electrons will localize and at what energy level.

"The power of the landscape function is letting you govern the waves, letting you design the system in which you can actually control the localization, [rather than letting] it be given by the gods," said Mayboroda.

And that, as it turns out, is exactly what you need to build a better LED.

ORDER AND LIGHT

LEDs are often hailed as the future of lighting. They're much better at efficiently converting energy into light than conventional bulbs. But LEDs are still a bit like a found resource: We've got this thing, we know it's useful, but we don't completely understand how to make it better.

"You lack control in this situation. You don't know why you've performed well, and you don't know what to do to go even further," Filoche said.

What we do know is that LEDs work through localization. LEDs contain thin layers of semiconducting materials bounded by electrodes. Those electrodes apply a voltage that sets the electrons in motion. The electrons move by hopping from one atom to another, assuming new positions in the "potential" energy map as they do so. As the electrons move, they leave behind positively charged "holes" that interact with electrons in important ways. As for the electrons themselves, when they move from positions of higher energy to positions of lower energy, under the right circumstances they emit the difference as a photon of light. Concentrate enough of these photons and you can banish the dark.

Of course, electrons don't always go where you want them to. Modern LEDs are made from wafers of a semiconducting alloy, gallium nitride, which surround even thinner layers of a related alloy, indium gallium nitride. These thin interior layers are evocatively called a "quantum well"—when electrons fall in, they localize at lower energy levels. If they localize in the presence of a hole, the energy difference is emitted as a photon of light; if they localize without a hole, the energy difference is emitted as a photon of heat and the whole effort is for naught.

So that's the setup: You want electrons to localize in quantum wells in the presence of holes to emit light. For a number of reasons, gallium nitride is a good material in which to make this happen, but it also has drawbacks—due to the way it's manufactured, you end up with a material that's very irregular at the atomic level.

"You will find regions of space where you have more indium atoms, and other regions with less indium atoms. This random change of composition means energies of electrons in different regions are different," said Claude

Weisbuch, a leading figure in semiconductor physics at the University of California, Santa Barbara, and co-recipient (with James Speck, also of UCSB) of a grant from the U.S. Department of Energy to use the landscape function to develop better green LEDs.

The landscape function maps the potential energy in the messy materials used to make LEDs. It tells you where electron waves will interfere to cancel each other out, and where electrons will localize, and at what energies. For engineers trying to make these devices, it's like turning on a bright light in a dark room.

"For the first time we can do real quantum simulations of LEDs thanks to landscape theory," Weisbuch said.

Mayboroda finished the first version of the landscape function six years ago. Since that time it has sprawled into a number of different research areas: At MIT, Jerison is exploring the broader mathematical implications of the function; in France, Filoche is using a scanning tunneling microscope to experimentally assess the function's predictions, and a separate research team (led by Patrick Sebbah of the Langevin Institute) is directly measuring localizations in vibrating plates; and in California, Weisbuch is designing new LEDs. Altogether, it's a stunning pace of application.

"It is amazing to me what has happened in a couple years," Mayboroda said. "I don't believe it myself."

PENTAGON TILING PROOF SOLVES CENTURY-OLD MATH PROBLEM

Natalie Wolchover

One of the oldest problems in geometry asks which shapes tile the plane, locking together with copies of themselves to cover a flat area in an endless pattern called a tessellation. M. C. Escher's drawings of tessellating lizards and other creatures illustrate that an unlimited variety of shapes can do this. The inventorying reduces to a finite, though still formidable, task when mathematicians consider only convex polygons: simple, flat-edged shapes like triangles and rectangles whose angles all bend in the same direction. In 2017, a proof by Michaël Rao, a mathematician at CNRS (France's national center for scientific research) and the École Normale Supérieure de Lyon, finally completed the classification of convex polygons that tile the plane by conquering the last holdouts: pentagons, which had resisted sorting for 99 years.

Try placing regular pentagons—those with equal angles and sides—edge to edge and gaps soon form; they do not tile. The ancient Greeks proved that the only regular polygons that tile are triangles, quadrilaterals and hexagons (as now seen on many a bathroom floor). But squash and stretch a pentagon into an irregular shape and tilings become possible. In his 1918 doctoral thesis, the German mathematician Karl Reinhardt identified five types of irregular convex pentagons that tile the plane: They were families defined by common rules, such as "side a equals side b," "c equals d" and "angles A and C both equal 90 degrees."

Reinhardt didn't know whether his five families completed the list, and progress stalled for 50 years. Then, in 1968, Richard Kershner of Johns Hopkins University discovered three more types of tessellating convex pentagons and claimed to have proved that no others existed.[1] But Kershner's paper left out the proof that his list was exhaustive "for the excellent reason," reads an introductory note, "that a complete proof would require a rather large book."

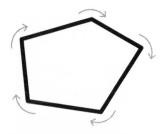

CONVEX PENTAGON

As you travel around the shape,
you always turn in the same direction.

CONCAVE PENTAGON

As you travel around the shape,
you turn both right and left.

FIGURE 3.8

News of Kershner's pentagon claim spread to the masses in 1975 when it appeared in Martin Gardner's popular math column in *Scientific American*. But soon after, lay readers like Marjorie Rice, a San Diego housewife with a high school math education, discovered new tessellating pentagon families beyond those known to Kershner. (Rice found four and a computer programmer named Richard James found one.) The list of families grew to 13 and, in 1985, to 14. Then, in 2015, Casey Mann, an associate professor of mathematics at the University of Washington, Bothell, and collaborators used a computer search to discover a 15th type of tessellating convex pentagon.[2]

When Rao heard about Mann and his team's discovery, he set out to do an exhaustive search that would complete the classification of tessellating convex pentagons once and for all.

In his computer-assisted proof, Rao identified 371 possible scenarios for how corners of pentagons might come together in a tiling, and then he checked them all.[3] In the end, his algorithm determined that only the 15 known pentagon families can do it. His proof closes the field of convex polygons that tile the plane at 15 pentagons, three types of hexagons—all identified by Reinhardt in his 1918 thesis—and all quadrilaterals and triangles. (No tilings by convex polygons with seven or more sides exist.)

Mann said he and his collaborators had been working on taking a partial step toward an exhaustive proof when they heard the news from France. "Rao beat us to the punch," he said, adding wryly, "Which is great, because it saves us a lot of work."

Thomas Hales, a professor of mathematics at the University of Pittsburgh and a leader in using computer programming to solve problems in geometry, has independently reproduced the most important half of Rao's proof, indicating that there isn't a bug.

As one journey—the classification of all convex polygon tessellations—ends, another is just beginning. Rao, like many tiling specialists, seeks the elusive "einstein," a hypothetical shape that can only tile the plane non-periodically, in a pattern of tile orientations that never repeats. Its name has nothing to do with the famous physicist but rather is German for "one stone." "For everybody who works on tiling, this is a kind of holy grail," Rao said. He sees his pentagon proof as an early milestone on this much larger quest.

GOOD SETS

In his proof, Rao first showed that there are only a finite number of scenarios for how the corners of convex pentagons can fit together that need to be checked for tessellations. He used simple geometric conservation laws to impose restrictions on how a pentagon's corners—labeled 1 to 5—can possibly meet at the vertices in a tiling. These conditions include the fact that the sum of angles 1 to 5 must equal 540 degrees—the total for any pentagon—and that all five have to participate in a tiling equally, since they're all part of every pentagonal tile. Moreover, the sums of the angles at a given vertex must always equal either 360 degrees, if the corners of the adjacent pentagons all meet there, or 180 degrees, if some corners meet along another pentagon's edge.

By imposing such rules, Rao found that other than for 371 scenarios, "either the angle equations or the percentages [dictating how often different angles appear] self-contradict," said Greg Kuperberg, a professor of mathematics at the University of California, Davis. A finite number of "good sets," as Rao dubbed the possible sets of angle conditions, wasn't guaranteed, Kuperberg said, "but his computer run delivered good news."

In the second of the two main steps in his proof, Rao went through the good sets one by one and checked whether any tilings satisfying these angle conditions existed. When it came to the coding, this was "the more complicated part," Rao said.

"For each of the 371 scenarios," Kuperberg explained, "his algorithm tries to piece together a tiling by laying down one tile at a time, using only the allowed vertex configurations." In searching through these 371 trees of possibilities, the algorithm either determined that every path in a tree leads to a pentagon in one of the 15 known families, or that "all paths on the tree lead to grief after a finite number of steps," explained Rich Schwartz, a mathematician at Brown University who works on related problems.

Rao said he felt disappointed not to have discovered any additional families, but tiling experts say that proving a complete list of 15 is more significant than simply finding a new working example.

The exhaustive proof also helps direct the search for the hypothetical einstein—that coveted puzzle piece that locks together with itself in an ever-changing sequence of tile orientations.

"One conclusion from Rao's breakthrough result is that there is no single convex polygon that tiles the plane nonperiodically," Kuperberg said. Since all 15 of the tessellating convex pentagons (and all other convex polygons) tile the plane periodically, meaning in a sequence of tile orientations that regularly repeats, the einstein, if it exists, must be concave, with jagged corners that bend both inward and outward like the corners of a star.

SEEKING EINSTEIN

Experts say there's good reason to think that the einstein exists, though its shape is probably very complex. That such an elusive shape would be needed to tessellate the plane nonperiodically only adds to its allure.

Nonperiodic tilings exist when you have tiles of at least two different shapes to play with—an example is the famous Penrose tiling—or when using a bizarre tile consisting of parts that are not connected, called the Socolar–Taylor tile. But whether a single connected tile exists that can do the job, and what its properties might be, remains unknown. Mann said the einstein's existence is "considered likely because it's connected to another very central problem in tiling theory" called the decision problem. "The question is, if someone hands you a tile, can you come up with a computer algorithm that will take as input that tile and say, 'Yes, this tiles the plane,' or, 'No, it doesn't'?"

"Most people think there's too much complexity for such an algorithm to exist," Mann said. Researchers have already proved that no algorithm exists that can decide if an arbitrary collection of different shapes tiles the plane. Many experts suspect, though it isn't proven, that the single-tile decision problem is "undecidable" as well. In a backward kind of way, this would imply the existence of the einstein tile. Since "it's pretty easy to check if something is periodic," Mann said, the decision problem should be decidable if single shapes only ever tile the plane periodically. The existence of the einstein tile and the hardness of the single-tile decision problem go hand in hand.

Rao plans to put his algorithms on the einstein's trail, though he says concave shapes represent a much bigger combinatorial problem than convex ones. Other experts were glad to hear he's on the case. Not long ago,

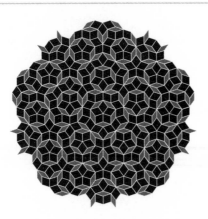

PENROSE TILING

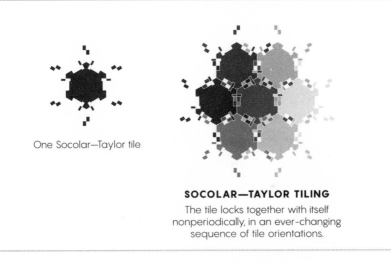

One Socolar—Taylor tile

SOCOLAR—TAYLOR TILING
The tile locks together with itself
nonperiodically, in an ever-changing
sequence of tile orientations.

FIGURE 3.9
Source: Inductiveload (top) and Joshua Socolar and Joan Taylor (bottom)

Rao and a collaborator proved a different result about nonperiodic tilings of Wang tiles—squares whose colored edges can only be placed side by side if the colors match.[4] Previous work had demonstrated that collections of Wang tiles exist that only give rise to nonperiodic tilings. "The first one that was found had over 20,000 tiles in it," Mann said. "That later got reduced to 14. Rao proved that you can do it with 11 tiles and that's the minimum. So he kind of knocked that problem out of consideration, too. He's on a tear."

SIMPLE SET GAME PROOF STUNS MATHEMATICIANS

Erica Klarreich

I n a series of papers posted online in 2016, mathematicians solved a problem about the pattern-matching card game Set that predates the game itself. The solution, whose simplicity has stunned mathematicians, is already leading to advances in other combinatorics problems.

Invented in 1974, Set has a simple goal: to find special triples called "sets" within a deck of 81 cards. Each card displays a different design with four attributes—color (which can be red, purple or green), shape (oval, diamond or squiggle), shading (solid, striped or outlined) and number (one, two or three copies of the shape). In typical play, 12 cards are placed face-up and the players search for a set: three cards whose designs, for each attribute, are either all the same or all different.

Occasionally, there's no set to be found among the 12 cards, so the players add three more cards. Even less frequently, there's still no set to be found among the 15 cards. How big, one might wonder, is the largest collection of cards that contains no set?

The answer is 20—proved in 1971 by the Italian mathematician Giuseppe Pellegrino.[1] But for mathematicians, this answer was just the beginning. After all, there's nothing special about having designs with only four attributes—that choice simply creates a manageable deck size. It's easy to imagine cards with more attributes (for instance, they could have additional images, or even play different sounds or have scratch-and-sniff smells). For every whole number n, there's a version of Set with n attributes and 3^n different cards.

For each such version, we can consider collections of cards that contain no set—what mathematicians confusingly call "cap sets"—and ask how large they can be. Mathematicians have calculated the maximal size of cap sets for games with up to six attributes, but we'll probably never know the exact size of the largest cap set for a game with 100 or 200 attributes, said

Jordan Ellenberg, a mathematician at the University of Wisconsin, Madison—there are so many different collections of cards to consider that the computations are too mammoth ever to be carried out.

Yet mathematicians can still try to figure out an upper bound on how big a cap set can be—a number of cards guaranteed to hold at least one set. This question is one of the simplest problems in the mathematical field called Ramsey theory, which studies how large a collection of objects can grow before patterns emerge.

"The cap set problem we think of as a model problem for all these other questions in Ramsey theory," said Terence Tao of UCLA. "It was always believed that progress would come there first, and then once we'd sorted that out we would be able to make progress elsewhere."

Yet until recently, this progress has been slow. Mathematicians established in papers published in 1995 and 2012 that cap sets must be smaller than about $1/n$ the size of the full deck.[2] Many mathematicians wondered, however, whether the true bound on cap set size might be much smaller than that.

They were right to wonder. The papers posted online in 2016 showed that relative to the size of the deck, cap set size shrinks exponentially as n gets larger. In a game with 200 attributes, for instance, the previous best result limited cap set size to at most about 0.5 percent of the deck; the new bound shows that cap sets are smaller than 0.0000043 percent of the deck.

The previous results were "already considered to be quite a big breakthrough, but this completely smashes the bounds that they achieved," said Timothy Gowers, a mathematician and Fields medalist at the University of Cambridge.

There's still room to improve the bound on cap sets, but in the near term, at least, any further progress is likely to be incremental, Gowers said. "In a certain sense this completely finishes the problem."

GAME, SET, MATCH

To find an upper bound on the size of cap sets, mathematicians translate the game into geometry. For the traditional Set game, each card can be encoded as a point with four coordinates, where each coordinate can take one of three values (traditionally written as 0, 1 and 2). For instance, the card with two striped red ovals might correspond to the point (0, 2, 1, 0), where the 0 in the first spot tells us that the design is red, the 2 in the second spot tells us that the shape is oval and so on. There are similar encodings for versions of Set with n attributes, in which the points have n coordinates instead of four.

The rules of the Set game translate neatly into the geometry of the resulting n-dimensional space: Every line in the space contains exactly three points, and three points form a set precisely when they lie on the same line. A cap set, therefore, is a collection of points that contains no complete lines.

Previous approaches to getting an upper bound on cap set size used a technique called Fourier analysis, which views the collection of points in a cap set as a combination of waves and looks for the directions in which the collection oscillates. "The conventional wisdom was that this was the way to go," Tao said.

This time, however, mathematicians solved the cap set problem using an entirely different method—and in only a few pages of fairly elementary mathematics. "One of the delightful aspects of the whole story to me is that I could just sit down, and in half an hour I had understood the proof," Gowers said.

The proof uses the "polynomial method," an innovation that, despite its simplicity, only rose to prominence on the mathematical scene about a decade ago. The approach produces "beautiful short proofs," Tao said. It's "sort of magical."

A polynomial is a mathematical expression built out of numbers and variables raised to powers—for instance, $x^2 + y^2$ or $3xyz^3 + 2$. Given any collection of numbers, it's possible to create a polynomial that evaluates to zero at all those numbers—for example, if you pick the numbers 2 and 3, you can build the expression $(x-2)(x-3)$; this multiplies out to the polynomial $x^2 - 5x + 6$, which equals zero if $x=2$ or $x=3$. Something similar can be done to create polynomials that evaluate to zero at a collection of points— for instance, the points corresponding to Set cards.

At first glance, this doesn't seem like a very deep fact. Yet somehow, these polynomials often seem to contain information that isn't readily visible from the set of points. Mathematicians don't fully understand, Ellenberg said, just why this approach works so well, and which types of problems it can be useful for. Until the papers appeared online, he added, he considered cap set "an example of a problem where the polynomial method really has no purchase."

That changed on May 5, 2016, when three mathematicians—Ernie Croot of the Georgia Institute of Technology, Vsevolod Lev of the University of Haifa, Oranim, in Israel and Péter Pál Pach of the Budapest University

Set Point

Set is played with cards bearing four attributes—shape (diamond, oval or squiggle), number (one, two or three copies of the shape), color (represented here as light, medium or dark gray) and shading (solid, striped or outlined). Players search for sets of three cards whose designs, for each attribute, are either all the same or all different. A collection of cards with no set is called a cap set.

THE FULL DECK

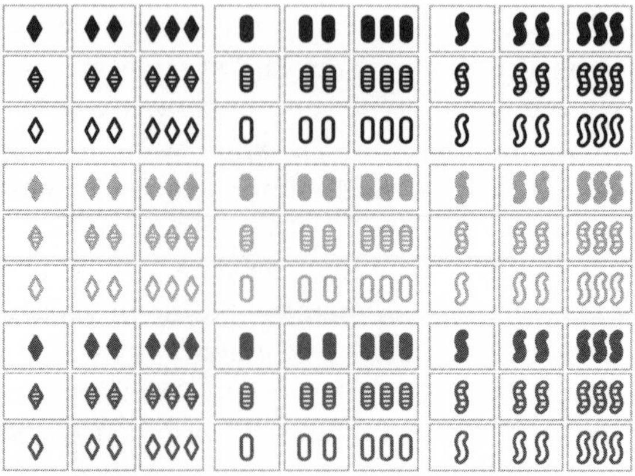

CAP SET

A simple way to build a fairly large cap set is to include only cards that show two of the three choices for each attribute. This cap set will be $(2/3)^n$ as big as the whole deck, where n is the number of attributes.

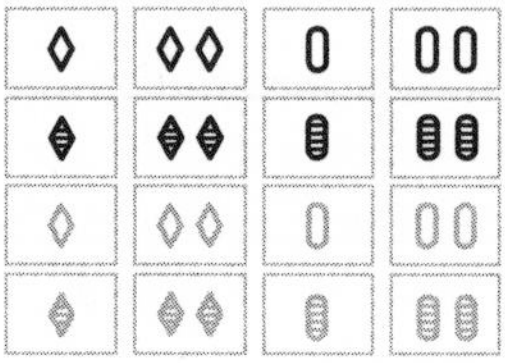

SET OR NO SET

Some examples:

Are the attributes all the same or all different?			
Color	✗	all different	all the same
Shape	all different	all different	all the same
Shading	all different	all different	all the same
Number	✗	all different	all different
	NOT A SET	A SET	A SET

FIGURE 3.10

of Technology and Economics in Hungary—posted a paper online show-ing how to use the polynomial method to solve a closely related problem, in which each Set attribute can have four different options instead of three.[3] For technical reasons, this problem is more tractable than the original Set problem.

In this game variant, for any collection of cards with no set, Croot, Lev and Pach considered which additional cards could be laid down on the table to complete a set. They then built a polynomial that evaluates to zero on these completion cards, and figured out an ingeniously simple way to split the polynomial into pieces with smaller exponents, which led to a bound on the size of collections with no sets. It was a "very inventive move," Ellenberg said. "It's always incredibly cool when there's something truly new and it's easy."

The paper soon set off a cascade of what Ellenberg called "math at inter-net speed." Within 10 days, Ellenberg and Dion Gijswijt, a mathematician at Delft University of Technology in the Netherlands, had each indepen-dently posted papers (later jointly published in the *Annals of Mathematics*) showing how to modify the argument to polish off the original cap set problem in just three pages.[4] The trick, Ellenberg said, is to realize that there are many different polynomials that evaluate to zero on a given set of points, and that choosing just the right one gets "a little bit more juice out of the method." A cap set, the new proofs establish, can be at most $(2.756/3)^n$ as large as the whole deck.

Mathematicians quickly scrambled to figure out the implications of the new proof. Soon, a paper was posted online showing that the proof rules out one of the approaches mathematicians were using to try to create more efficient matrix multiplication algorithms.[5] And that same month, Gil Kalai, of the Hebrew University of Jerusalem, wrote an "emergency" blog post pointing out that the cap set result can be used to prove the "Erdős–Szemerédi sunflower conjecture," which concerns sets that overlap in a sun-flower pattern.

"I think a lot of people will be thinking, 'What can I do with this?'" Gowers said. Croot, Lev and Pach's approach, he wrote in a blog post, is "a major new technique to add to the toolbox."

The fact that the cap set problem finally yielded to such a simple tech-nique is humbling, Ellenberg said. "It makes you wonder what else is actu-ally easy."

A MAGICAL ANSWER TO AN 80-YEAR-OLD PUZZLE

Erica Klarreich

In 2015, the mathematician Terence Tao of UCLA presented a solution to an 80-year-old number theory problem posed by the legendary Hungarian mathematician Paul Erdős. Erdős was famous for the thousands of puzzles he came up with, many of which have led to surprisingly deep mathematical discoveries. This particular problem, which came to be known as the Erdős discrepancy problem, was one of his favorites, said Ben Green, a mathematician at the University of Oxford. "He mentioned it many times over the years, particularly towards the end of his life."

A simplified version of the problem goes like this: Imagine that you are imprisoned in a tunnel that opens out onto a precipice two paces to your left, and a pit of vipers two paces to your right. To torment you, your evil captor forces you to take a series of steps to the left and right. You need to devise a series that will allow you to avoid the hazards—if you take a step to the right, for example, you'll want your second step to be to the left, to avoid falling off the cliff. You might try alternating right and left steps, but here's the catch: You have to list your planned steps ahead of time, and your captor might have you take every second step on your list (starting at the second step), or every third step (starting at the third), or some other skip-counting sequence. Is there a list of steps that will keep you alive, no matter what sequence your captor chooses?

In this brainteaser, devised by the mathematics popularizer James Grime, you can plan a list of 11 steps that protects you from death. But if you try to add a 12th step, you are doomed: Your captor will inevitably be able to find some skip-counting sequence that will plunge you over the cliff or into the viper pit.

Around 1932, Erdős asked, in essence, what if the precipice and pit of vipers are three paces away instead of two? What if they are N paces away? Can you escape death for an infinite number of steps? The answer, Erdős conjectured, was no—no matter how far away the precipice and viper pit are, you can't elude them forever.

But for more than 80 years, mathematicians made no progress on proving Erdős' discrepancy conjecture (so named because the distance from the center of the tunnel is known as the discrepancy). "Everyone in the subject has whetted their teeth on this and failed," said Andrew Granville of the University of Montreal and University College London. "It's one of those problems that nobody has really written a sensible paper about, because no one had a clever idea."

Even the seemingly simple scenario in which the pit and the precipice are just three paces away involves an enormous number of possible choices. This version of the problem was finally solved in 2014 by Boris Konev and Alexei Lisitsa of the University of Liverpool in England, who showed, via a computer calculation whose output is comparable in size with all of Wikipedia, that it is possible to write down 1,160 safe steps, but no more.[1] Their proof, however, did not offer a foothold on the more general problem.

Then, in a preprint posted online on September 17, 2015, Tao showed that no matter how far away the viper pit and precipice are, there is always a maximum number of steps you can safely list.[2]

To solve the problem, Tao measured the "entropy" of mathematical objects called multiplicative functions or sequences, which lie at the heart of not just the Erdős discrepancy problem, but also some of the deepest problems in number theory, such as understanding the distribution of prime numbers. This novel synthesis of number theory and entropy—a concept that originated in coding theory—"will certainly open up new avenues of research," Green predicted.

Until this result, the Erdős discrepancy problem had been an example of "the most ridiculous things we have felt we didn't understand" about multiplicative functions, Granville said. "It should be an immediate observation, but somehow it had to take vast amounts of deep ideas and cleverness to get there." Tao's solution, he said, is "a wonderful breakthrough."

HANDLES ON THE WRECKING BALL

In late 2009, Timothy Gowers, a mathematician at the University of Cambridge who jump-started the massive online mathematical collaborations known as "Polymath" projects, was casting about for a good topic for the next such project. In a series of blog posts, he described several possible projects, including the Erdős discrepancy problem, and asked readers to weigh in. The post on the discrepancy problem quickly attracted nearly 150 comments, and on January 6, 2010, Gowers wrote what he called an "emergency" post saying that this problem was clearly the people's choice.

Like Erdős himself, the project cast the problem as a question about sequences of +1s and −1s, not rights and lefts. Over the course of the project, Tao figured out that it is essentially sufficient to solve the discrepancy problem for multiplicative sequences: ones in which the $(n \times m)$th entry is equal to the nth entry times the mth entry (so, for example, the sixth entry equals the second entry times the third entry).

It makes sense that multiplicative sequences should offer high prospects for survival. In a multiplicative sequence, each skip-counting sequence of +1s and −1s is either identical to or the mirror image of the original sequence as a whole. For instance, the sequence that consists of every third entry is simply the original sequence times the third entry in the sequence, which is either +1 or −1. So, if you've found a survivable list of steps for the main sequence, it will automatically give you a survivable list of steps for every skip-counting sequence your captor might choose.

Multiplicative sequences are related to deep structures in number theory. One example is the famous Liouville function, which, when written as a sequence, has a +1 or −1 in the nth spot depending on whether n has an even or odd number of prime factors, and which gives mathematicians a way to study the number of primes below a given number. Multiplicative sequences have been intensively studied, but many basic questions about them have stubbornly resisted attempts to answer them. One such question, the Polymath project eventually concluded, was the Erdős discrepancy problem. "By 2012, [the Polymath project] petered out," Tao said.

But the problem somehow seemed more approachable after the Polymath project, he said. Before, the problem had been "like a giant wrecking ball that you had to pick up, but it was completely smooth," he said, using an analogy he attributes to Gowers. After Polymath, "the problem had handles," he said. "So you could at least try to pick it up now. If you found a crane, you could hook it up."

HOISTING THE WRECKING BALL

In January 2015, a pair of mathematicians—Kaisa Matomäki of the University of Turku in Finland and Maksym Radziwiłł of McGill University— took the first step toward building that crane, although it wasn't immediately clear that they had done so. They came up with a way to understand the correlations between near neighbors in a multiplicative sequence, an achievement that had long been considered beyond reach.[3]

Tao started working with Matomäki and Radziwiłł on a raft of potential applications of their method to problems in number theory. On September 6 that year, Tao wrote a blog post about some of this work pertaining to the Liouville function, and he mentioned that the problem reminded him of a Sudoku puzzle. A few days later, a Polymath participant named Uwe Stroinski commented that the Erdős discrepancy problem also had a Sudoku-like flavor. Could Matomäki and Radziwiłł's approach, he asked, be applied to that problem as well?

"I replied saying, 'No, I don't think so,'" Tao said. He was convinced—as in fact proved to be the case—that every sequence eventually leads to death in the Erdős puzzle. Matomäki and Radziwiłł's approach seemed as if it might be useful for constructing sequences that allow you to survive for a while, but not for the reverse problem of showing that the sequence must eventually fail. As Tao gave the question more thought, however, he realized that his knee-jerk response was wrong—he could in fact prove the Erdős conjecture, if he could only control a certain complicated sum.

"I tried seriously to tackle this head-on, now that I knew that it would solve the discrepancy problem," Tao said. And one afternoon, as he waited for his son to get out of a piano lesson, the answer came to him: He could use an argument "like a magician's choice, where the magician offers someone in the audience two options, and it seems as if the audience has control, but the magician has a trick planned for whichever option you pick."

Tao's trick involves breaking up a candidate sequence into chunks and then examining the sequence, chunk by chunk, to see whether you can survive your captor. When you encounter a new chunk, one of two things must happen, Tao showed: Either the captor can kill you, or the sequence's entropy—a measure of how random a sequence is—will drop by a definite increment. Entropy can never drop below zero, so if you continue on from chunk to chunk, you must eventually hit a chunk for which the only possibility is that your captor can kill you.

Tao solved the problem in the space of a month, which is "an amazing testament to his strength," Granville said. "Once he gets his teeth into something, he can't let go."

Tao, who was 10 when he first met Erdős at a mathematics event, is excited about the power of the "magician's trick" approach, he said. "I hope it can be used to prove many other things."

SPHERE PACKING SOLVED IN HIGHER DIMENSIONS

Erica Klarreich

I n a pair of papers posted online in 2016, a Ukrainian mathematician solved two high-dimensional versions of the centuries-old "sphere packing" problem. In dimensions eight and 24 (the latter dimension in collaboration with other researchers), she proved that two highly symmetrical arrangements pack spheres together in the densest possible way.

Mathematicians have been studying sphere packings since at least 1611, when Johannes Kepler conjectured that the densest way to pack together equal-sized spheres in space is the familiar pyramidal piling of oranges seen in grocery stores. Despite the problem's seeming simplicity, it was not settled until 1998, when Thomas Hales, now of the University of Pittsburgh, finally proved Kepler's conjecture in 250 pages of mathematical arguments combined with mammoth computer calculations.[1]

Higher-dimensional sphere packings are hard to visualize, but they are eminently practical objects: Dense sphere packings are intimately related to the error-correcting codes used by cell phones, space probes and the internet to send signals through noisy channels. A high-dimensional sphere is easy to define—it's simply the set of points in the high-dimensional space that are a fixed distance away from a given center point.

Finding the best packing of equal-sized spheres in a high-dimensional space should be even more complicated than the three-dimensional case Hales solved, since each added dimension means more possible packings to consider. Yet mathematicians have long known that two dimensions are special: In dimensions eight and 24, there exist dazzlingly symmetric sphere packings called E_8 and the Leech lattice, respectively, that pack spheres better than the best candidates known to mathematicians in other dimensions.

"Somehow everything just fits perfectly together, and it's sort of a miracle," said Henry Cohn, a mathematician at Microsoft Research New England in Cambridge, Massachusetts. "I don't have a simple, gut-level explanation of what it is about."

For reasons that mathematicians don't fully understand, E_8 and the Leech lattice have connections to a wide range of mathematical subjects, including number theory, combinatorics, hyperbolic geometry and even areas of physics such as string theory. They form "a nexus where lots of different areas of mathematics come together," Cohn said. "Something wonderful is happening, and I'd like to understand what it is."

Mathematicians had amassed compelling numerical evidence that E_8 and the Leech lattice are the best sphere packings in their respective dimensions. But that evidence had come just short of a proof. Researchers had known for more than a decade what the missing ingredient in the proof should be—an "auxiliary" function that can calculate the largest allowable sphere density—but they couldn't find the right function.

Then, in a paper posted online on March 14, 2016, Maryna Viazovska came up with the missing function in dimension eight.[2] Her work used the theory of modular forms, powerful mathematical functions that, when they can be brought to bear upon a problem, seem to unlock huge amounts of information. In this case, finding the right modular form allowed Viazovska, then a postdoctoral researcher at the Berlin Mathematical School and the Humboldt University of Berlin, to prove, in a mere 23 pages, that E_8 is the best eight-dimensional packing.

"It's stunningly simple, as all great things are," said Peter Sarnak, of Princeton University and the Institute for Advanced Study. "You just start reading the paper and you know this is correct."

Within a week, Viazovska, along with Cohn and three other mathematicians, successfully extended her method to cover the Leech lattice too.

"I think some of us have been hoping for this for a very long time," Hales said.

FILLING THE GAPS

It's possible to build an analogue of the pyramidal orange stacking in every dimension, but as the dimensions get higher, the gaps between the high-dimensional oranges grow. By dimension eight, these gaps are large enough to hold new oranges, and in this dimension only, the added oranges lock tightly into place. The resulting eight-dimensional sphere packing, known as E_8, has a much more uniform structure than its two-stage construction might suggest. "Part of the mystery here is this object turns out to be vastly more beautiful and symmetric than it sounds," Cohn said. "There are tons of extra symmetries."

The Leech lattice is similarly constructed by adding spheres to a less dense packing, and it was discovered almost as an afterthought. In the 1960s, the British mathematician John Leech was studying a 24-dimensional packing that can be constructed from the "Golay" code, an error-correcting code that was later used to transmit the historic photos of Jupiter and Saturn taken by the Voyager probes. Shortly after Leech's article about this packing went to press, he noticed that there was room to fit additional spheres into the holes in the packing, and that doing so would double the packing density.[3]

In the resulting Leech lattice, each sphere is surrounded by 196,560 other spheres, in such a unique arrangement that the mathematician John Conway, of Princeton University, discovered three entirely new types of symmetry by probing the lattice's structure.[4] The Leech lattice is "one of the few most exciting mathematical objects," said Gil Kalai, a mathematician at the Hebrew University of Jerusalem.

In 2003, Cohn and Noam Elkies of Harvard University developed a way to estimate just how well E_8 and the Leech lattice perform compared to other sphere packings in their respective dimensions.[5] In every dimension, Cohn and Elkies showed, there is an infinite sequence of "auxiliary" functions that can be used to compute upper limits on how dense sphere packings are allowed to be in that dimension.

In most dimensions, the best sphere packings discovered to date didn't even come close to the density limits this method generated. But Cohn and Elkies found that in dimensions eight and 24, the best packings—E_8 and the Leech lattice—seemed to practically bump their heads against the ceiling. When Cohn and Abhinav Kumar of Stony Brook University carried out extensive numerical calculations on the sequences of auxiliary functions, they found that the best possible sphere packings in dimensions eight and 24 could be at most 0.0000000000000000000000001 percent denser than E_8 and the Leech lattice.[6]

Given this ridiculously close estimate, it seemed clear that E_8 and the Leech lattice must be the best sphere packings in their respective dimensions. Cohn and Elkies suspected that for each of these two dimensions, there should be some auxiliary function that would give an exact answer that matched the density of E_8 and the Leech lattice. "We gave many talks and even convened a conference or two to disseminate the problem in the hope that such [a function] was known or could easily be found if we only know in which mathematical field to look, but found nothing," Elkies wrote in an email.

Hales said he had believed for years that the right function should exist, but he had no idea how to pin it down. "I felt that it would take a Ramanujan to find it," he said, referring to the early 20th-century mathematician

Srinivasa Ramanujan, who was famous for pulling deep mathematical ideas out of thin air.

Then Viazovska found the elusive auxiliary functions for E_8 and the Leech lattice, using a type of mathematical object that Ramanujan also studied extensively: modular forms. "She's pulled a Ramanujan," Hales said.

MINING FOR GOLD

Modular forms are functions that possess special symmetries like those in M. C. Escher's circular tilings of angels and devils. These functions possess a startling power to illuminate different areas of mathematics—for instance, they were instrumental in the proof of Fermat's Last Theorem in 1994. And although modular forms have been studied for centuries, mathematicians are still unlocking the deep secrets hidden inside their coefficients. Sarnak calls them a gold mine. "I'm waiting for someone to write a paper one day, 'The Unreasonable Effectiveness of Modular Forms,'" he said.

Unfortunately, though, there's only a limited supply of modular forms, and they are highly constrained objects. "You can't just write down a modular form that does whatever you want," Cohn said. "So it's a matter of whether one actually exists that does what you need it to."

Viazovska's 2013 doctoral dissertation was on modular forms, and she also has expertise in discrete optimization, one of the fields that are central to the sphere-packing problem. So when, five years ago, Viazovska's friend Andrii Bondarenko, of the Norwegian University of Science and Technology in Trondheim, suggested that they work together on the eight-dimensional sphere-packing problem, Viazovska agreed.

They worked on the problem off and on along with Danylo Radchenko of the Max Planck Institute for Mathematics in Germany. Eventually, Bondarenko and Radchenko moved on to other problems, but Viazovska pressed on alone. "I felt like it's my problem," she said.

After two years of intense effort, she succeeded in coming up with the right auxiliary function for E_8 and proving that it is correct. It's difficult, she said, to explain just how she knew which modular form to use, and she's currently writing a paper to try to describe her "philosophical reason" for searching for it where she did. There's "a whole new mathematical story behind it," she said.

After Viazovska posted her paper on March 14, 2016, she was startled by the surge of excitement it created among sphere-packing researchers. "I thought people would be interested in the result, but I did not know there would be that much attention," she said.

That night, Cohn emailed to congratulate her, and as the two exchanged emails he asked if it might be possible to extend her method to the Leech lattice. "I felt like, 'I am already tired and I deserve some rest,'" Viazovska said. "But I tried still to be useful."

The two of them threw together a collaboration with Kumar, Radchenko and Stephen Miller of Rutgers University, and with the benefit of Viazovska's earlier work, they quickly found a way to construct the right auxiliary function for the Leech lattice. The team posted its 12-page paper online just a week after Viazovska had posted her first paper.[7]

The results don't have practical implications for error-correcting codes, since knowing that E_8 and the Leech lattice were close to perfect had already been sufficient for all real-world applications. But the two proofs offer mathematicians both a sense of closure and a powerful new tool. A natural next question, Cohn said, is whether these methods can be adapted to show that E_8 and the Leech lattice have "universal optimality." This would mean that they provide not just the best sphere packings but also the lowest-energy ones, if, for example, the centers of the spheres are regarded as electrons repelling one another.

And because E_8 and the Leech lattice are connected to so many areas of mathematics and physics, Viazovska's approach may well ultimately lead to many more discoveries, said Akshay Venkatesh of Stanford University. "It seems to me much more likely than not that this function is also part of some richer story."

IV HOW DO THE BEST MATHEMATICAL MINDS WORK?

A TENACIOUS EXPLORER OF ABSTRACT SURFACES

Erica Klarreich

A s an 8-year-old, Maryam Mirzakhani used to tell herself stories about the exploits of a remarkable girl. Every night at bedtime, her heroine would become mayor, travel the world or fulfill some other grand destiny.

It was August 2014, and the 37-year-old Stanford mathematician was still writing elaborate stories in her mind. The high ambitions hadn't changed, but the protagonists had become hyperbolic surfaces, moduli spaces and dynamical systems. In a way, she said, mathematics research felt like writing a novel. "There are different characters, and you are getting to know them better," she told *Quanta* at the time. "Things evolve, and then you look back at a character, and it's completely different from your first impression."

The Iranian mathematician followed her characters wherever they took her, along story lines that often took years to unfold. Petite but indomitable, Mirzakhani had a reputation among mathematicians for tackling the most difficult questions in her field with dogged persistence. With her low voice and steady, gray-blue eyes, Mirzakhani projected an unwavering self-confidence. She had an equal tendency, however, toward humility. When asked to describe her contribution to a particular research problem, she laughed, hesitated and finally said: "To be honest, I don't think I've had a very huge contribution." And when an email arrived in February 2014 saying that she would receive what is widely regarded as the highest honor in mathematics—the Fields Medal—she assumed that the account from which the email was sent had been hacked.

Other mathematicians, however, described Mirzakhani's work in glowing terms. Her doctoral dissertation—about counting loops on surfaces that have "hyperbolic" geometry—was "truly spectacular," said Alex Eskin, a mathematician at the University of Chicago who has collaborated with Mirzakhani. "It's the kind of mathematics you immediately recognize belongs in a textbook."

And Mirzakhani's monumental collaboration with Eskin about the dynamics of abstract surfaces connected to billiard tables is "probably the theorem of the decade" in Mirzakhani's highly competitive field, said Benson Farb, also a University of Chicago mathematician.[1]

TEHRAN

As a child growing up in Tehran, Mirzakhani had no intention of becoming a mathematician. Her chief goal was simply to read every book she could find. She also watched television biographies of famous women such as Marie Curie and Helen Keller, and later read *Lust for Life*, a novel about Vincent van Gogh. These stories instilled in her an undefined ambition to do something great with her life—become a writer, perhaps.

Mirzakhani finished elementary school just as the Iran–Iraq war was drawing to a close and opportunities were opening up for motivated students. She took a placement test that secured her a spot at the Farzanegan middle school for girls in Tehran, which is administered by Iran's National Organization for Development of Exceptional Talents. "I think I was the lucky generation," she told *Quanta*. "I was a teenager when things got more stable."

In her first week at the new school, she made a lifelong friend, Roya Beheshti, who is now a mathematics professor at Washington University in St. Louis. As children, the two explored the bookstores that lined the crowded commercial street near their school. Browsing was discouraged, so they randomly chose books to buy. "Now, it sounds very strange," Mirzakhani said, "but books were very cheap, so we would just buy them."

To her dismay, Mirzakhani did poorly in her mathematics class that year. Her math teacher didn't think she was particularly talented, which undermined her confidence. At that age, "it's so important what others see in you," Mirzakhani said. "I lost my interest in math."

The following year, Mirzakhani had a more encouraging teacher, however, and her performance improved enormously. "Starting from the second year, she was a star," Beheshti said.

Mirzakhani went on to the Farzanegan high school for girls. There, she and Beheshti got hold of the questions from that year's national competition to determine which high school students would go to the International Olympiad in Informatics, an annual programming competition for high school students. Mirzakhani and Beheshti worked on the problems for several days and managed to solve three out of six. Even though students at the competition must complete the exam in three hours, Mirzakhani was excited to be able to do any problems at all.

Eager to discover what they were capable of in similar competitions, Mirzakhani and Beheshti went to the principal of their school and demanded that she arrange for math problem-solving classes like the ones being taught at the comparable high school for boys. "The principal of the school was a very strong character," Mirzakhani recalled. "If we really wanted something, she would make it happen." The principal was undeterred by the fact that Iran's International Mathematical Olympiad team had never fielded a girl, Mirzakhani said. "Her mindset was very positive and upbeat—that 'you can do it, even though you'll be the first one,'" Mirzakhani said. "I think that has influenced my life quite a lot."

In 1994, when Mirzakhani was 17, she and Beheshti made the Iranian math Olympiad team. Mirzakhani's score on the Olympiad test earned her a gold medal. The following year, she returned and achieved a perfect score. Having entered the competitions to discover what she could do, Mirzakhani emerged with a deep love of mathematics. "You have to spend some energy and effort to see the beauty of math," she said.

Twenty years later, Anton Zorich of the Université Paris Diderot-Paris 7 in France, said Mirzakhani still gave "the impression of a 17-year-old girl who is absolutely excited by all the mathematics that happens around her."

HARVARD

Gold medals at the mathematical Olympiad don't always translate into success in mathematics research, said Curtis McMullen of Harvard University, who was Mirzakhani's doctoral adviser. "In these contests, someone has carefully crafted a problem with a clever solution, but in research, maybe the problem doesn't have a solution at all." Unlike many Olympiad high-scorers, he said, Mirzakhani had "the ability to generate her own vision."

After completing an undergraduate degree in mathematics at Sharif University of Technology in Tehran in 1999, Mirzakhani went to graduate school at Harvard University, where she started attending McMullen's seminar. At first, she didn't understand much of what he was talking about but was captivated by the beauty of the subject, hyperbolic geometry. She started going to McMullen's office and peppering him with questions, scribbling down notes in Farsi.

"She had a sort of daring imagination," recalled McMullen, a 1998 Fields medalist. "She would formulate in her mind an imaginary picture of what must be going on, then come to my office and describe it. At the end, she would turn to me and say, 'Is it right?' I was always very flattered that she thought I would know."

Mirzakhani became fascinated with hyperbolic surfaces—doughnut-shaped surfaces with two or more holes that have a non-standard geometry which, roughly speaking, gives each point on the surface a saddle shape. Hyperbolic doughnuts can't be constructed in ordinary space; they exist in an abstract sense, in which distances and angles are measured according to a particular set of equations. An imaginary creature living on a surface governed by such equations would experience each point as a saddle point.

It turns out that each many-holed doughnut can be given a hyperbolic structure in infinitely many ways—with fat doughnut rings, narrow ones or any combination of the two. In the century and a half since such hyperbolic surfaces were discovered, they have become some of the central objects in geometry, with connections to many branches of mathematics and even physics.

But when Mirzakhani started graduate school, some of the simplest questions about such surfaces were unanswered. One concerned straight lines, or "geodesics," on a hyperbolic surface. Even a curved surface can have a notion of a "straight" line segment: it's simply the shortest path between two points. On a hyperbolic surface, some geodesics are infinitely long, like straight lines in the plane, but others close up into a loop, like the great circles on a sphere.

The number of closed geodesics of a given length on a hyperbolic surface grows exponentially as the length of the geodesics grows. Most of these geodesics cut across themselves many times before closing up smoothly, but a tiny proportion of them, called "simple" geodesics, never intersect themselves. Simple geodesics are "the key object to unlocking the structure and geometry of the whole surface," Farb said.

Yet mathematicians couldn't pin down just how many simple closed geodesics of a given length a hyperbolic surface can have. Among closed geodesic loops, the simple ones are "miracles that [effectively] happen zero percent of the time," Farb said. For that reason, counting them accurately is incredibly difficult: "If you have a little bit of an error, you've missed it," he said.

In her doctoral thesis, completed in 2004, Mirzakhani answered this question, developing a formula for how the number of simple geodesics of length L grows as L gets larger. Along the way, she built connections to two other major research questions, solving both. One concerned a formula for the volume of the so-called "moduli" space—the set of all possible hyperbolic structures on a given surface. The other was a surprising new proof of an old conjecture proposed by the physicist Edward Witten of the Institute for Advanced Study, about certain topological measurements of moduli spaces related to string theory. Witten's conjecture is so difficult that the first mathematician to prove it—Maxim Kontsevich of the Institut des Hautes Études Scientifiques, near Paris—was awarded a Fields Medal in 1998 in part for that work.

Farb said that solving each of these problems "would have been an event, and connecting them would have been an event." Mirzakhani did both. Mirzakhani's thesis resulted in three papers published in the three top journals of mathematics: *Annals of Mathematics, Inventiones Mathematicae* and *Journal of the American Mathematical Society*.[2] The majority of mathematicians will never produce something as good, Farb said—"and that's what she did in her thesis."

"A TITANIC WORK"

Mirzakhani liked to describe herself as slow. Unlike some mathematicians who solve problems with quicksilver brilliance, she gravitated toward deep problems that she could chew on for years. "Months or years later, you see very different aspects" of a problem, she said. There are problems she thought about for more than a decade without finding the answer.

Mirzakhani didn't feel intimidated by mathematicians who knock down one problem after another. "I don't get easily disappointed," she said. "I'm quite confident, in some sense."

Her slow and steady approach also applied to other areas of her life. One day while she was a graduate student at Harvard, her future husband, then a graduate student at the Massachusetts Institute of Technology, learned this lesson about Mirzakhani when the two went for a run. "She's very petite, and I was in good shape, so I thought I'd do well, and at first, I was ahead," recalled Jan Vondrák, who is now an associate professor at Stanford University. "But she never slows down. After half an hour, I was done, but she was still running at the same pace."

As she thought about mathematics, Mirzakhani constantly doodled, drawing surfaces and other images related to her research. "She has these huge pieces of paper on the floor and spends hours and hours drawing what look to me like the same picture over and over," Vondrák told *Quanta* in 2014, adding that papers and books were scattered haphazardly about her home office. "I have no idea how she can work like this, but it works out in the end," he said. Perhaps, he speculated, "the problems she is working on are so abstract and complicated, she can't afford to make logical steps one by one but has to make big jumps."

Doodling helped her focus, Mirzakhani said. When thinking about a difficult math problem, "you don't want to write down all the details," she said. "But the process of drawing something helps you somehow to stay connected." Mirzakhani said that her daughter, Anahita, who was then 3, often exclaimed, "Oh, Mommy is painting again!" when she saw the mathematician drawing. "Maybe she thinks I'm a painter," Mirzakhani said.

Mirzakhani's research connects to many areas of mathematics, including differential geometry, complex analysis and dynamical systems. "I like crossing the imaginary boundaries people set up between different fields— it's very refreshing," she said. In her area of research, "there are lots of tools, and you don't know which one would work," she said. "It's about being optimistic and trying to connect things."

Sometimes, the connections that Mirzakhani made were mind-blowing, McMullen said. In 2006, for example, she tackled the problem of what happens to a hyperbolic surface when its geometry is deformed using a mechanism akin to a strike-slip earthquake. Before Mirzakhani's work, "this problem was completely unapproachable," McMullen said. But with a one-line proof, he said, "she constructed a bridge between this completely opaque theory and another theory that's completely transparent."

In 2006, Mirzakhani began her fruitful collaboration with Eskin, who considered her one of his favorite collaborators. "She is very optimistic, and that's infectious," he said in 2014. "When you work with her, you feel you have a much better chance of solving problems that at first seem hopeless."

After several projects together, Mirzakhani and Eskin decided to tackle one of the largest open problems in their field. It concerned the range of behaviors of a ball that is bouncing around a billiard table shaped like any polygon, provided the angles are a rational number of degrees. Billiards provides some of the simplest examples of dynamical systems—systems that evolve over time according to a given set of rules—but the behavior of the ball has proven unexpectedly hard to pin down.

"Rational billiards got started a century ago, when some physicists were sitting around saying, 'Let's understand a billiard ball bouncing in a triangle,'" said Alex Wright, then a postdoctoral researcher at Stanford. "Presumably, they thought they would be done in a week, but 100 years later, we're still thinking about it."

To study a long billiard ball trajectory, a useful approach is to imagine gradually deforming the billiard table by squishing it along the direction of the trajectory so that more of the ball's path can be seen in a given amount of time. This transforms the original billiard table into a succession of new ones, moving the table around in what mathematicians call the "moduli" space consisting of all possible billiard tables with a given number of sides. By transforming each billiard table into an abstract surface called a "translation surface," mathematicians can analyze billiard dynamics by understanding the larger moduli space consisting of all translation surfaces. Researchers have shown that understanding the "orbit" of a particular translation surface as the squishing action moves it around in the moduli space helps in answering a host of questions about the original billiard table.

On the face of it, this orbit might be an extremely complicated object—a fractal, for instance. In 2003, however, McMullen showed that this isn't the case when the translation surface is a two-holed ("genus two") doughnut: Every single orbit fills up either the entire space or some simple subset of the space called a submanifold.

McMullen's result was hailed as a huge advance. He recalled that before his paper was published, however, Mirzakhani—then still a graduate student—came to his office and asked, "Why did you just do genus two?"

"That's the kind of person she is," he said. "What she sees hints of, she wants to understand more clearly."

After years of work, in 2012 and 2013, Mirzakhani and Eskin, partly in collaboration with Amir Mohammadi now of the University of California, San Diego, succeeded in generalizing McMullen's result to all doughnut surfaces with more than two holes.[3] Their analysis was "a titanic work," Zorich said, adding that its implications go far beyond billiards. The moduli space "has been intensively studied for the last 30 years," he said, "but there's still so much we don't know about its geometry."

Mirzakhani and Eskin's work was "the beginning of a new era," said Wright, who spent months studying their 172-page paper.[4] "It's as if we were trying to log a redwood forest with a hatchet before, but now they've invented a chain saw," he said. Their work has already been applied—for example, to the problem of understanding the sightlines of a security guard in a complex of mirrored rooms.[5]

In Mirzakhani and Eskin's paper, "under every layer of difficulty and ideas lay another, hidden beneath," Wright wrote in an email. "By the time I got to the center, I was amazed at the machine they had built."

It was Mirzakhani's optimism and tenacity that kept the pair going, Eskin said. "Sometimes there were setbacks, but she never panicked," he said.

Even Mirzakhani herself was amazed, in retrospect, that the two stuck with it. "If we knew things would be so complicated, I think we would have given up," she said. Then she paused. "I don't know; actually, I don't know," she said. "I don't give up easily."

A HISTORIC FIRST

Mirzakhani was the first woman to win a Fields Medal. The gender imbalance in mathematics is long-standing and pervasive, and the Fields Medal, in particular, is ill-suited to the career arcs of many female mathematicians. It is restricted to mathematicians younger than 40, focusing on the very years during which many women dial back their careers to raise children.

Mirzakhani felt certain, however, that there will be many more female Fields medalists in the future. "There are really many great female mathematicians doing great things," she said.

But while she felt greatly honored to have been awarded a Fields Medal, she had no desire to be the face of women in mathematics, she said. Her ambitious teenage self would have been overjoyed by the award, she said, but she was eager to deflect attention from her achievements so she could focus on research.

Mirzakhani had big plans for the next chapters of her mathematical story. She started working with Wright to try to develop a complete list of the kinds of sets that translation surface orbits can fill up. Such a classification would be a "magic wand" for understanding billiards and translation surfaces, Zorich has written.[6]

It was no small task, but Mirzakhani had learned over the years to think big. "You have to ignore low-hanging fruit, which is a little tricky," she told *Quanta*. "I'm not sure if it's the best way of doing things, actually—you're torturing yourself along the way." But she enjoyed it, she said. "Life isn't supposed to be easy."

On July 14, 2017, Mirzakhani died of cancer at the age of 40.

A "REBEL" WITHOUT A PH.D.

Thomas Lin

F reeman Dyson—the world-renowned mathematical physicist who helped found quantum electrodynamics with the bongo-playing, Nobel Prize-winning physicist Richard Feynman and others, devised numerous mathematical techniques, led the team that designed a low-power nuclear reactor that produces medical isotopes for research hospitals, dreamed of exploring the solar system in spaceships propelled by nuclear bombs, wrote technical and popular science books, penned dozens of reviews for the *New York Review of Books* and turned 90 in December 2013—was pondering a new math problem.

"There's a class of problem that Freeman just lights up on," said the physicist and computational biologist William Press, a longtime colleague and friend. "It has to be unsolved and well-posed and have something in it that admits to his particular kind of genius." That genius, he said, represents a kind of "ingenuity and a spark" that most physicists lack: "The ability to see further in the mathematical world of concepts and instantly grasp a path to the distant horizon that's the solution."

Press said he'd posed a number of problems to Dyson that didn't "measure up." Months and years went by, with no response. But when Press asked a question about the "iterated prisoner's dilemma," a variation of the classic game theory scenario pitting cooperation against betrayal, Dyson replied the next day. "It probably only took him a minute to grasp the solution," Press said, "and half an hour to write it out."

Together, they published a much-cited 2012 paper in the *Proceedings of the National Academy of Sciences*.[1]

The next year, Press traveled to Princeton, New Jersey, for a two-day celebration of Dyson at the Institute for Advanced Study, Dyson's intellectual home for the past six decades. In honor of Dyson's 90th birthday, there was seemingly boundless cake, a forest of long, white candles, 350 guests—including his 16 grandchildren—and lectures recognizing his eclectic achievements in math, physics, astronomy and public affairs. H. T. Yau of Harvard University commenced the math section, launching into Dyson's work on the universality of random matrices. George Andrews of Pennsylvania State University and Kathrin Bringmann of the University of Cologne followed with the implications of Dyson's early contributions to number theory, which he began contemplating in high school. William Happer, a physicist at Princeton University and a fellow skeptic of the perils of anthropogenic climate change, closed day one with a talk provocatively titled "Why Has Global Warming Paused?"

Dyson admitted to being controversial when it comes to climate science. But during an hour-long interview with *Quanta Magazine* just days after that birthday celebration, he said: "Generally speaking, I'm much more of a conformist." Still, he has written fondly of science as an act of rebellion. In his 2006 anthology of essays and reviews, *The Scientist as Rebel*, Dyson wrote, "I was lucky to be introduced to science at school as a subversive activity of the younger boys."[2] With characteristic concern for social issues, he went on to advise parents: "We should try to introduce our children to science today as a rebellion against poverty and ugliness and militarism and economic injustice."

On the second day of the 2013 celebration in Princeton, after numerous speakers had recounted past collaborations with Dyson, alternately feting and roasting his brilliance, Press took a different tack. Referring to their collaboration on the prisoner's dilemma, Press—a professor at the University of Texas, Austin—said he "thought it would be a little extreme to reminisce with Freeman about a paper that was just published." Instead, he described his own recent result on safer "adaptive" clinical trials, adding that although he had solid computational data, the mathematical analysis proved too formidable.[3] "I wish I had worked on it with Freeman—and maybe still will get the chance to do so," he said slyly.

Press' comment proved prescient. After the celebration, Dyson began mulling over the problem—unbeknownst to Press, who didn't find out until *Quanta* contacted him in March 2014 about the new "collaboration." "I'm glad to know it's on his stack of things to do!" he said. "I'm looking forward to seeing what he comes up with."

Quanta Magazine interviewed Dyson at the institute in December 2013. The interview was edited and condensed for clarity.

Technically, you retired from the Institute for Advanced Study 20 years ago. What are you working on now?

I used to be a scientist and did a lot of calculations. It was a competitive world, and when I got older, I decided I wouldn't compete with the bright, young people anymore, so I write books instead. And now I've become a book reviewer for the *New York Review of Books*. About once a month, I write a review, and then I get a lot of response and correspondence, people who are finding things I said which aren't true.

What did you do prior to writing book reviews?
I was trained as a mathematician, and I remain a mathematician. That's really my skill, just doing calculations and applying mathematics to all kinds of problems, and that led me into physics first and also other fields, such as engineering and even a bit of biology, sometimes a little bit of chemistry. Mathematics applies to all kinds of things. That's one of the joys of being a mathematician.

Why math?
I think the decisive moment was reading the book *Men of Mathematics* by Eric Temple Bell. Bell was a professor at Caltech, and he wrote this book, which is actually just a wonderful collection of biographies of mathematicians. Historians condemn it as romanticized. But what was wonderful about this book is that he showed the mathematicians as being mostly crooks and people of very mixed kinds of qualities, not at all saints, and many of them quite unscrupulous and not very clever, and still they managed to do great mathematics. So it told a kid that "if they can do it, why can't you?"

What are some of the big questions that have guided your career?
I'm not a person for big questions. I look for puzzles. I look for interesting problems that I can solve. I don't care whether they're important or not, and so I'm definitely not obsessed with solving some big mystery. That's not my style.

What kinds of puzzles first intrigued you?
I started out as a pure mathematician and found problems that just arise out of the very nature of numbers, which are amazingly subtle and difficult and beautiful. That was when I was about 17 or so, just at the end of high school. I was interested in numbers before I was interested in the real world.

What is it about numbers that made you want to figure them out?
It's just like asking, "Why does a violinist like to play the violin?" I had this skill with mathematical tools, and I played these tools as well as I could just because it was beautiful, rather in the same way a musician plays the violin, not expecting to change the world but just because he loves the instrument.

You're known for your work in quantum electrodynamics—which describes interactions between light, matter and charged particles—and in solving the renormalization problem—which helped rid the mathematics of unwanted infinities. How did that work come about?

When I arrived in Cornell in 1947, there just had been done a beautiful experiment at Columbia on the hydrogen atom. The hydrogen atom is the simplest atom, and you ought to be able to understand it if you understand atoms at all. So, these experiments were done by Willis Lamb and his student Robert Retherford at Columbia, observing for the first time the very fine behavior of hydrogen using microwaves to examine the hydrogen atoms, and Lamb got very precise results. The problem was the quantum theory wasn't good enough to explain his results. Dick Feynman, who was an absolute genius, had understood more or less how to explain it but couldn't translate his ideas into ordinary mathematics. I came along and had the mathematical skill, making it possible to calculate precisely what the hydrogen atom was doing, and the amazing thing was that my calculations all agreed with the experiment, so it turns out the theory was right.

I didn't invent anything new—I translated Feynman's ideas into mathematics so it became more accessible to the world, and, as a result, I became famous, but it all happened within about six months.

Did it lead to other questions that you wanted to explore?

I got job offers from everywhere in America and also in England, but the problem was that I didn't actually want to settle down yet and become an overburdened professor with lots of students. So I escaped to England and had two happy years at Birmingham without any responsibilities and continued working on other problems.

I was very much interested in space travel, and so the next exciting thing I did was to work with a company in California called General Atomics for a couple of years building a spaceship. In those days, people were willing to take all kinds of risks, and all kinds of crazy schemes got supported. So there was this bunch of crazy, young people—the leader was Freddie de Hoffmann, who had been at Los Alamos [National Laboratory] and knew all about nuclear bombs—and we decided we would go around the solar system with a spaceship driven by nuclear bombs. We would launch the ship into space—"bomb, bomb, bomb, bomb," about four bombs per second—going up all the way to Mars and then afterwards to Jupiter and Saturn, and we intended to go ourselves.

What happened to Project Orion?
I spent two wonderful years in San Diego having grand dreams of space-ships. We not only did calculations, we also flew little models about a meter in diameter with chemical explosives, which actually went "bomb, bomb, bomb, bomb" a few times a few hundred feet up. It was amazing we never got hurt. I think we didn't even have to buy the explosives. We had some Navy friend who stole it from the Navy. Anyhow, we certainly borrowed the test stand from the Navy where we did these little flight tests. That lasted for two years. By that time, it was clear that the competition was actually going to win, the competition being Wernher von Braun and the Apollo program, which was going to go with ordinary rockets to the moon.

The Orion spaceship sounds like something a child might dream up. How disappointed were you that this "grand dream" wasn't realized?
Of course we were very disappointed when it turned out that the Orion never flew, but it was clear that it would make a horrible mess of the land-scape. These bombs were producing radioactive fallout as they went up through the atmosphere, and although at that time we were exploding bombs in the atmosphere for military purposes, which were much bigger than the ones we proposed to use, still we would have made a contribution to the general contamination, and that was the reason why the project failed, and I think it was a good reason.

You've developed a reputation as a maverick scientist with contrarian views. Where do you think that comes from?
I think the notion that I always like to oppose the consensus in science is totally wrong. The fact is there's only one subject that I've been controver-sial on, which is climate. I spend maybe 1 percent of my time on climate, and that's the only field in which I'm opposed to the majority. Generally speaking, I'm much more of a conformist, but it happens I have strong views about climate because I think the majority is badly wrong, and you have to make sure if the majority is saying something that they're not talk-ing nonsense.

With a majority of scientists on the other side of this issue, what would it take to convince you to switch sides?
What I'm convinced of is that we don't understand climate, and so that's sort of a neutral position. I'm not saying the majority is necessarily wrong. I'm saying that they don't understand what they're seeing. It will take a lot

of very hard work before that question is settled, so I shall remain neutral until something very different happens.

You became a professor at Cornell without ever having received a Ph.D. You seem almost proud of that fact.

Oh, yes. I'm very proud of not having a Ph.D. I think the Ph.D. system is an abomination. It was invented as a system for educating German professors in the 19th century, and it works well under those conditions. It's good for a very small number of people who are going to spend their lives being professors. But it has become now a kind of union card that you have to have in order to have a job, whether it's being a professor or other things, and it's quite inappropriate for that. It forces people to waste years and years of their lives sort of pretending to do research for which they're not at all well suited. In the end, they have this piece of paper which says they're qualified, but it really doesn't mean anything. The Ph.D. takes far too long and discourages women from becoming scientists, which I consider a great tragedy. So I have opposed it all my life without any success at all.

How is it that you were able to escape that requirement?

I was lucky because I got educated in World War II and everything was screwed up so that I could get through without a Ph.D. and finish up as a professor. Now that's quite impossible. So, I'm very proud that I don't have a Ph.D. and I raised six children and none of them has a Ph.D., so that's my contribution.

Looking back at your career, how has your approach to science changed over the decades?

I've now been active for something like 70 years, and still I use the same mathematics. I think the main thing that's changed as a result of computers is the magnitude of databases. We now have these huge amounts of data and very little understanding. So what we have now—I forget who it was who said this—are small islands of understanding in a sea of information. The problem is to enlarge the islands of understanding.

What scientific advance do you see on the horizon that will have a big impact on society?

People are often asking me what's going to happen next in science that's important, and of course, the whole point is that if it's important, it's something we didn't expect. All the really important things come as a big surprise. There are many examples of this, of course, dark energy being the latest example. Anything I mention will be something that, obviously, is not a surprise.

Are you currently working on a math problem?

The question of what I do with my time is a delicate one. I'm not really doing science competitively, but I like to have a problem to work on. I'm very lucky to have a friend, Bill Press, who is an expert on clinical trials, which actually turns out to be an interesting mathematical problem.

He published a paper explaining how to do clinical trials in a really effective way with a minimum loss of life. He's a computer expert, so everything he does is worked out just with numbers, and so I have taken on as my next task to translate what he did into equations, the same way I did with Feynman. I'm not sure whether it will work, but that's what I'm thinking about at the moment.

What does it mean for someone with so many intellectual pursuits to be retired?

When I retired as a professor of the institute, I kept all the privileges. The only thing that changed is the paychecks stopped coming. I still have an office and all the secretarial help I need, plus a place at the lunch table. One more advantage is not having to go to faculty meetings.

A BRAZILIAN WUNDERKIND WHO CALMS CHAOS

Thomas Lin and Erica Klarreich

I t was pouring rain on a chilly spring day in 2014, and Artur Avila was marooned at the University of Paris Jussieu campus, minus the jacket he had misplaced before boarding a red-eye from Chicago. "Let's wait," said the Brazilian mathematician in a sleep-deprived drawl, his snug black T-shirt revealing the approximate physique of a sturdy World Cup mid-fielder. "I don't want to get sick." In everyday matters, Avila steers clear of complications and risk. Afraid his mind will veer from road signs and oncoming traffic to "unimodal maps" and "quasi-periodic Schrödinger operators," he doesn't drive or bike. "There are too many cars in Paris," he said. "I'm fearful of some crazy bus killing me."

Soon the conversation turned to a different kind of worry for Avila—that public reminders of Brazil's apparent lack of intellectual achievement will discourage students there from pursuing careers in pure math and science research. In the lead-up to the World Cup competition that year, popular news websites and TV shows like "Good Morning Brazil" parroted the question: How has the world's seventh-largest economy managed to score five World Cup titles but zero Nobel Prizes? (The British biologist Peter Medawar's tenuous connection to Brazil—born there but raised in his mother's native England—merits at best an asterisk.) Even Argentina, that bitter soccer rival with a population one-fifth the size of Brazil's, boasts five Nobel laureates.

To Avila, the criticism stung. "It's not good for the self-image of Brazil," he said.

Even then, months before it would be announced, the native son of Rio de Janeiro had a secret weapon, a compelling argument that Brazil belonged among elite math nations like the United States, France and Russia. But he could tell no one—not until August 2014, when the International Mathematical Union made Avila the first Brazilian recipient of the Fields Medal, awarding the 35-year-old what many consider the equivalent of a Nobel Prize in mathematics for his "profound contributions to dynamical systems theory" that "have changed the face of the field," according to the prize selection committee.

"He's one of the very best analysts in the world," said Jean-Christophe Yoccoz, a renowned Collège de France mathematician and 1994 Fields medalist, in 2014. Of the many talented postdoctoral researchers Yoccoz had advised, he said, "Artur is in a class by himself." Most mathematicians focus on a narrow subfield and have a low success rate, Yoccoz explained, but Avila "attacks many important problems and solves many of them."

His work "cannot be reduced to 'one big theorem' as Artur has so many deep results in several different topics," said Marcelo Viana, who worked with Avila to solve a long-standing problem about the chaotic behavior of billiard balls. The two proved a formula that predicts which side of the table a ball is most likely to hit next—and which side it will likely hit after a thousand bounces, or a million, all with the same margin of error. By contrast, Viana observed, if you try to predict the weather, "you'll get very good predictions for tomorrow, not very good predictions for the day after, and completely lousy predictions for 15 days from now."

Months before the Fields Medal announcement, the Brazilian dynamicist Welington de Melo predicted that his former doctoral student would win math's highest honor. "It's going to be extremely important to Brazil," he said. "We never before got such a high prize. It is especially important because Artur was a student in Brazil all the time."

MATH ON THE BEACH

Two things Avila fears more than erratic buses are PowerPoint slides and income tax forms. The pressure to perfect a plenary talk for the thousands attending the 2010 math congress in Hyderabad, India, induced in him a kind of mental paralysis, he said. After giving a lecture at the California Institute of Technology in 2008, he declined an honorarium of more than $2,000 just to avoid the paperwork.

"I would get fired pretty fast from most jobs," he said, adding that he sleeps well past noon and is "not good at managing time."

But in mathematics, Avila has a reputation for diving headfirst into unfamiliar waters and rapidly solving a raft of ambitious open questions. His colleagues describe his working style as highly collaborative and freakishly fast and Avila himself as having a clear-minded intuition for simplifying deep complications.

"He has high geometric vision," said Raphael Krikorian, an Armenian-French dynamicist at University of Cergy-Pontoise in France. "He tells you what you should look at, what you should do. Then, of course, you have to work."

Now a globe-trotting dual citizen of Brazil and France, Avila spends half the year in Paris as a research director at CNRS, France's largest state-run science organization, and the other half in Rio as a fellow at IMPA, Brazil's national institute for pure and applied mathematics. (The Brazil-France connection is no coincidence; in the 1970s and 1980s, top young French mathematicians like Étienne Ghys and Yoccoz fulfilled their mandatory military service with a civil-service alternative: conducting research at IMPA.)

In balmy Rio during the summer and winter months, Avila mulls over problems while lying in bed or wandering Leblon Beach a block from his apartment. There, he has more time and freedom to think deeply about his work and to let ideas flow freely. "I don't believe that I can just hit my head on the wall and the solution will appear," he likes to say. He sometimes invites collaborators to Rio one at a time for what can only be described as an unconventional work experience.

"The last time I was in Rio, I specifically got a hotel near the beach so I could work with him," said Amie Wilkinson, a mathematician at the University of Chicago. After searching for Avila on a beach that was "packed shoulder-to-shoulder" with "oversexed cariocas" and returning to her hotel to try to call him, Wilkinson eventually found him "literally standing in the water," she said. "We met and worked up to our knees in water. It was totally crazy."

"If you work with Artur," she said, "you have to get into a bathing suit."

Avila was born to parents who could not envision their son growing up to become a pure mathematician—they had never heard of one—and wanted him to aim for a stable career as a bureaucrat. His father's formal education growing up in the rural Amazon didn't start until his teenage years, but by the time Artur was born, his father had become an accountant in a government reinsurance enterprise, able to provide a middle-class lifestyle in Rio for his family and buy math books for his quiet son, who early on was more interested in reading than imitating Pelé's bicycle kick. When Avila was 6, his mother—who still files his tax returns—enrolled him at Colégio de São Bento, a conservative Catholic school known for its academics and for the gold-plated, 16th-century São Bento Monastery. Two years later his parents separated. As the years passed, Avila increasingly focused on mathematics to the exclusion of almost everything else—he often did poorly in other subjects and was expelled after the eighth grade for refusing to take mandatory religion exams. He said he "left the school completely unprepared for normal social interaction."

Avila got his first taste of the wider mathematics community just before he was expelled in 1992 when Luiz Fabiano Pinheiro, a master teacher at São Bento affectionately known as "Fabiano," encouraged the 13-year-old prodigy to participate in the junior division of the prestigious Mathematical Olympiad competition. Avila was excited by problems he had never encountered but felt woefully unprepared. "For the first time, I felt I couldn't do anything," he said. The next year, after Fabiano helped him transfer to a new school, Avila won top honors at the state level. Two years later, he took gold at the International Mathematical Olympiad in Toronto.

"The first time I met Artur, I knew that he would be pre-eminent," Fabiano said in Portuguese as his ex-wife, Eliana Vianna, interpreted. "Artur was the best of all my students ever," said the retiree who taught for five decades.

Through the math competitions, Avila discovered IMPA, where Brazil held its Olympiad award ceremonies each year. There, he met prominent mathematicians like Carlos Gustavo Moreira and Nicolau Corção Saldanha, and while still technically in high school, he began studying graduate-level mathematics.

DYNAMICAL SYSTEMS

In Brazil, Avila could relish mathematics without the career pressures he might have faced in the United States. "It was better for me to study at IMPA than if I were at Princeton or Harvard," he said. "Growing up and being educated in Brazil was very positive for me."

A major focus at IMPA is dynamical systems, the branch of mathematics that studies systems that evolve over time according to some set of rules—a collection of planets moving around a star, for example, or a billiard ball bouncing around a table, or a population of organisms that grows or declines over time.

One reason that many young mathematicians are drawn to dynamical systems, several researchers said, is that the relatively new subject, unlike the ancient field of number theory, doesn't require a great deal of prior theoretical knowledge to begin solving problems. And dynamical systems are everywhere in math and nature. "It's like a glue that connects many other subjects," Krikorian said. Of the "two cultures of mathematics" described by the University of Cambridge mathematician and 1998 Fields medalist Timothy Gowers, there are theory-builders who create new mathematics and there are problem-solvers who analyze existing questions. Most dynamicists, said Yoccoz, including Avila and himself, are problem-solvers. "Both ways are necessary," he said.

In the decades preceding Avila's work, mathematicians made a profound discovery: To produce complex behavior, it isn't necessary to start with complex rules. Even simple rules when repeated again and again sometimes produce chaos: random-seeming, unpredictable behavior in which tiny changes in the starting conditions can produce dramatically different outcomes. One of the first simple systems in which chaotic behavior was discovered is the so-called "logistic" model of population growth, which gives a precise formulation for how a population will change from year to year. Avila arrived on the scene just in time to write what could be considered the final chapter of this story.

In nature, a small population often grows quickly because there is an abundance of resources; a larger population will grow more slowly or even decline as resources are stretched too thin. In 1838, the Belgian mathematician Pierre Verhulst captured this intuition in the logistic equation for population growth. The graph of the logistic equation is simply an upside-down parabola that rises quickly if the population is small but drops precipitously if the population is larger than the environment can sustain. As a population changes in time, it will move around on the parabola—a small population may become large the next year and a large one small.

Not all populations respond the same way, of course. The logistic equation encodes this diversity with a parameter r between 1 and 4 that controls the steepness of the parabola: Higher values of r correspond to populations that react in more abrupt, extreme ways to small changes. Populations with low values of r will tend toward an equilibrium point or perhaps bounce around between a few values from year to year. But for certain values of r larger than about 3.56995—a value called the "onset of chaos"—the system becomes completely unpredictable.

"It's just a parabola—something kids learn how to graph in school," Wilkinson observed. "Yet that simple picture has this insanely rich stuff going on."

Researchers had known for decades that beyond the onset of chaos, there were "islands of stability"—values of r greater than 3.56995 for which the population would tend, for example, to a three-year cycle or a seven-year cycle. In the late 1990s, Mikhail Lyubich of Stony Brook University in Long Island elucidated what happens outside of these islands: For almost every other parameter beyond the onset of chaos, the equation's behavior is "stochastic," exhibiting the unpredictable bouncing around that is the hallmark of chaos.[1]

Lyubich had recently completed his analysis of the logistic equation when he traveled to IMPA in 1998. While there, he met Avila and instantly put the shy 19-year-old at ease. "As a student, I was scared of making mistakes," Avila said. "He was very gentle and not scary at all."

Lyubich, de Melo and Avila decided to try to extend Lyubich's analysis of the behavior after the onset of chaos to a more general setting. In the mid-1970s, mathematicians discovered that the particular mix of cycles and chaos that the logistic equation exhibits seems to be a universal feature of every family of equations with the same basic shape as the upside-down parabola (called unimodal maps).[2] Scientists also uncovered this same combination of cycles and chaos in a wide variety of systems in fluid dynamics, chemistry and other areas of science. Researchers struggled, though, to ground their observations in formal mathematics.

The three mathematicians investigated what happens to a family of unimodal maps after the onset of chaos. They distilled the problem down to a particular question, which Avila then solved. His construction, Lyubich wrote in a survey of Avila's work in 2012, "is delicate and first looked too good to be true—but it worked, and it completed the argument."[3]

The proof showed that a wide class of unimodal families behaves just like the logistic family: After the onset of chaos, there are islands of stability, surrounded nearly entirely by parameters that give rise to stochastic behavior.

HAMMER AND NAILS

In the spring and fall months, when he is based at CNRS in Paris, Avila is a mathematical free electron, moving from institute to institute in search of "attractive" problems. "Sometimes beauty is found in the mathematical statement and sometimes in the use of mathematical tools," he said. "When they mix together in an unexpected way, then it's something that I want to be working on."

Many mathematicians are drawn to Avila because he "demystifies" complicated ideas, making them seem trivial, said Bassam Fayad, a CNRS colleague. "If you work with him, this experience changes your attitude toward mathematics. You kind of learn to do math without pain."

Once, while wandering the streets of Paris with Avila, Jairo Bochi mentioned that he was trying to use spheres to construct a proof that he had been working on for two months. "People usually take a long time to understand your problem," said Bochi, a mathematician from southern Brazil who shared an apartment with Avila when they studied at IMPA.

"He immediately could see what I was saying and made a suggestion: 'The spheres won't work; try a compact cylinder.'" It solved the problem, and in 2006, they published their result.[4]

When jumping into a new field, Avila prefers to learn by talking to other people, rather than reading past papers. "Several times, I have gone aggressively to a new problem without having a lot of background in the area," Avila said. "Maybe, since I don't know what other people have been trying to do, I avoid some dead ends. Once I get a good result, then I'm more motivated to learn about the field, to understand what I'm proving."

Central to the proof on unimodal maps was a powerful technique called renormalization that can sometimes convert one dynamical system into a new, related system by zooming in on a small part that behaves similarly to the system as a whole. Avila became a master of this technique. "He contributed in a deep way to understanding this phenomenon," Lyubich said.

Renormalization was Avila's hammer, and he started realizing that nails were lying around everywhere he looked. He carried out a series of further collaborations on unimodal dynamical systems, which, Lyubich wrote in the survey, "effectively closed up the field." But that was just the beginning. Avila began diving into other areas of dynamical systems, using renormalization to solve one important problem after another.

"Part of Avila's strength is that he is capable of working in all these different areas and, in a sense, unifying them," Lyubich said. "He selects an area that looks interesting, finds the right fundamental problem to work on, then goes after it and is basically unstoppable."

Avila has also studied the evolution of quantum states in physical systems governed by "quasi-periodic Schrödinger operators," which are crude models for quasicrystals, structures that have more order than a liquid but less than a crystal. For these quasi-periodic Schrödinger operators, Avila—working with Svetlana Jitomirskaya of the University of California, Irvine, and David Damanik of Rice University in Houston—"just came along and cleaned up," Wilkinson said. "He answered a ton of questions about them." One of these questions, concerning the energy states that electrons can take on in a particular model for a quasicrystal, had been known informally as the Ten Martini Problem because it was so difficult that the late Polish mathematician Mark Kac had promised 10 martinis to anyone who could solve it.[5] Avila has recently been extending this understanding to an entire family of quasi-periodic Schrödinger operators.

Avila has worked with Wilkinson and Sylvain Crovisier of the Université Paris-Sud in Orsay, France, to study a famous hypothesis of the 19th-century Austrian physicist Ludwig Boltzmann. Boltzmann proposed that a gas inside a box is "ergodic," meaning that the gas atoms will rapidly travel through all possible arrangements, rather than, for example, hanging out for extended periods in particular regions of the box. In recent work, Avila, Crovisier and Wilkinson have shown that in the case of mathematical models whose behavior is at least moderately smooth, Boltzmann's ergodic hypothesis is true, except possibly in the case of certain systems that are highly predictable, akin to a billiard ball bouncing around a table.[6]

Although the vast majority of his papers have been collaborations, it has been years since Avila's former doctoral adviser worked with him. "I think he's too fast for me," said de Melo. "You have to work very hard to try to keep up with him. He would be happy to do almost everything, but I want to make sure I'm also contributing."

BRAZILIAN SUPERSTAR

Jet-lagged and stubble-faced after his trans-Atlantic flight in May 2014, Avila looked more like a postdoc than a master dynamicist with a wide-ranging anthology of greatest hits. Having made important contributions to his field since he was 19 (and having earned his Ph.D. at 21), he has long defied expectations. When it comes to math, if not helter-skelter city traffic or tax filings, Avila's shy anxieties have been replaced by a relaxed confidence and unwavering determination.

Fayad, who has known Avila since the Brazilian wunderkind first arrived in Paris in 2001, spoke of his friend's growing drive and professionalism, whether he is attacking big math problems or researching weight-training techniques on the internet. "He's not very amateur in things," Fayad said. "Suppose he wants to eat chocolate. He will become a professional eater of chocolate."

For all his early accomplishments, Avila insists that he doesn't set goals for himself, preferring to let his work unfold naturally. "Most of the time when I accomplish something, it's not because I had a goal but because I was doing something that I wanted to do," he said. "I just want to keep enjoying doing math."

And he hopes his home country will share his enthusiasm. "These four years are naturally going to be a good period to develop math in Brazil," Avila said in 2014. In addition to Avila's Fields Medal win, Brazil hosted the 2017 International Mathematical Olympiad and the 2018 International Congress of Mathematicians, where the next Fields medalists were announced.

He hopes it is only the beginning of a transformative movement that will elevate assumptions about his country's intellectual promise.

In 2013, at a small bar in Paris that happened to be playing Brazilian music, Avila overheard a French mathematician tell his friends that Brazil "did not really have a math school." When Avila objected, the other mathematician made it clear that he was aware of IMPA.

"I was a bit annoyed and insisted there was high-level math coming from there," Avila said, to which the French mathematician replied, yes, there is that Brazilian superstar—Avila or something. "He was expecting someone older."

THE MUSICAL, MAGICAL NUMBER THEORIST

Erica Klarreich

F or Manjul Bhargava, the counting numbers don't simply line themselves up in a demure row. Instead, they take up positions in space—on the corners of a Rubik's Cube, or the two-dimensional layout of the Sanskrit alphabet, or a pile of oranges brought home from the supermarket. And they move through time, in the rhythms of a Sanskrit poem or a tabla drumming sequence.

Bhargava's mathematical tastes, formed in his earliest days, are infused with music and poetry. He approaches all three realms with the same goal, he says: "to express truths about ourselves and the world around us."

The soft-spoken, boyish mathematician could easily be mistaken for an undergraduate student. He projects a quiet friendliness that makes it easy to forget that he is widely considered one of the towering mathematical figures of his age. "He's very unpretentious," said Benedict Gross, a mathematician at Harvard University who has known Bhargava since the latter's undergraduate days. "He doesn't make a big deal of himself."

Yet the search for artistic truth and beauty has led Bhargava, a mathematics professor at Princeton University, to some of the most profound recent discoveries in number theory, the branch of mathematics that studies the relationships between whole numbers. In recent years, he has made great strides toward understanding the range of possible solutions to equations known as elliptic curves, which have bedeviled number theorists for more than a century.

"His work is better than world-class," said Ken Ono, a number theorist at Emory University in Atlanta. "It's epoch-making."

And in 2014, Bhargava was awarded the Fields Medal, widely viewed as the highest honor in mathematics.

Bhargava "lives in a wonderful, ethereal world of music and art," Gross said. "He floats above the normal concerns of daily life. All of us are in awe of the beauty of his work."

Bhargava "has his own perspective that is remarkably simple compared to others," said Andrew Granville, a number theorist at the University of Montreal. "Somehow, he extracts ideas that are completely new or are retwisted in a way that changes everything. But it all feels very natural and unforced—it's as if he found the right way to think."

MUSICIAN

From early childhood, Bhargava displayed a remarkable mathematical intuition. "Teach me more math!" he would badger his mother, Mira Bhargava, a mathematics professor at Hofstra University in Hempstead, New York. When he was 3 years old and a typical, rambunctious toddler, his mother found that the best way to keep him from bouncing off the walls was to ask him to add or multiply large numbers.

"It was the only way I could make him stay still," she recalled. "Instead of using paper and pencil, he would kind of flip his fingers back and forth and then give me the right answer. I always wondered how he did it, but he wouldn't tell me. Perhaps it was too intuitive to explain."

Bhargava saw mathematics everywhere he looked. At age 8, he became curious about the oranges he would stack into pyramids before they went into the family juicer. Could there be a general formula for the number of oranges in such a pyramid? After wrestling with this question for several months, he figured it out: If a side of a triangular pyramid has length n, the number of oranges in the pyramid is $n(n+1)(n+2) / 6$. "That was an exciting moment for me," he said. "I loved the predictive power of mathematics."

Bhargava quickly became bored with school and started asking his mother if he could go to work with her instead. "She was always very cool about it," he recalled. Bhargava explored the university library and went for walks in the arboretum. And, of course, he attended his mother's college-level math classes. In her probability class, the 8-year-old would correct his mother if she made a mistake. "The students really enjoyed that," Mira Bhargava said.

Every few years, Bhargava's mother took him to visit his grandparents in Jaipur, India. His grandfather, Purushottam Lal Bhargava, was the head of the Sanskrit department of the University of Rajasthan, and Manjul Bhargava grew up reading ancient mathematics and Sanskrit poetry texts.

To his delight, he discovered that the rhythms of Sanskrit poetry are highly mathematical. Bhargava is fond of explaining to his students that the ancient Sanskrit poets figured out the number of different rhythms with a given number of beats that can be constructed using combinations of long and short syllables: It's the corresponding number in what Western mathematicians call the Fibonacci sequence. Even the Sanskrit alphabet has an inherent mathematical structure, Bhargava discovered: Its first 25 consonants form a 5 by 5 array in which one dimension specifies the bodily organ where the sound originates and the other dimension specifies a quality of modulation. "The mathematical aspect excited me," he said.

At Bhargava's request, his mother started teaching him to play tabla, a percussion instrument of two hand drums, when he was 3 (he also plays the sitar, guitar and violin). "I liked the intricacy of the rhythms," he said, which are closely related to the rhythms in Sanskrit poetry. Bhargava eventually became an accomplished player, even studying tabla with the legendary Zakir Hussain in California. He has performed in concert halls around the country and even at Central Park in New York City.

"He's a terrific musician who has reached a very high technical level," said Daniel Trueman, a music professor at Princeton who collaborated with Bhargava on a performance over the internet with musicians in Montreal. Just as important, he said, is Bhargava's warmth and openness. Even though Trueman's background is not primarily in Indian music, "I never felt that I was offending his high level of knowledge of North Indian classical music," Trueman said.

Bhargava often turns to the tabla when he is stuck on a mathematics problem, and vice versa. "When I go back, my mind has cleared," he said.

He experiences playing the tabla and doing mathematics research similarly, he said. Indian classical music—like number theory research—is largely improvisational. "There's some problem-solving, but you're also trying to say something artistic," he said. "It's similar to math—you have to put together a sequence of ideas that enlightens you."

Mathematics, music and poetry together feel like a very complete experience, Bhargava said. "All kinds of creative thoughts come together when I think about all three."

MATHEMATICIAN

Between attending his mother's classes and traveling to India, Bhargava missed a lot of school over the years. But on the days he didn't go to school, he would often meet his schoolmates in the afternoon to play tennis and basketball. Despite his extraordinary intelligence, "he was just a normal kid, associating with all the kids," Mira Bhargava recalled. "They were completely at ease with him."

That's a refrain repeated by Bhargava's colleagues, students and fellow musicians, who describe him using words like "sweet," "charming," "unassuming," "humble" and "approachable." Bhargava wears his mathematical superstardom lightly, said Hidayat Husain Khan, a professional sitarist based in Princeton and India who has performed with him. "He has the ability to connect with a huge spectrum of people, regardless of their background."

The only time that Bhargava's extended school absences threatened to harm him was when his high school health teacher tried to block him from graduating—even though he was the valedictorian and had been accepted to Harvard. (He did graduate.)

It was at Harvard that Bhargava decided, once and for all, to pursue a career in mathematics. With such eclectic interests, he had flirted with many possible careers—musician, economist, linguist, even mountain climber. Eventually, however, he realized that it was usually the mathematical aspects of these subjects that got him most excited.

"Somehow, I always came back to math," he said.

Bhargava felt the strongest tug between mathematics and music but decided in the end that it would be easier to be a mathematician who did music on the side than a musician who did mathematics on the side. "In academia, you can pursue your passions," he said.

Now, Bhargava has an office on the 12th floor of Princeton's Fine Hall littered with math toys—Rubik's Cubes, Zometools, pine cones and puzzles. When he is thinking about mathematics, however, Bhargava prefers to escape his office and wander in the woods. "Most of the time when I'm doing math, it's going on in my head," he said. "It's inspirational being in nature."

This approach can have its drawbacks: More than once, Bhargava has postponed writing down an idea for years only to forget the specifics. At times, however, delays between thinking and writing are inevitable. "Sometimes, when I have a new idea, there hasn't been language developed to express it yet," he said. "Sometimes, it's just a picture in my mind of how things should flow."

Although Bhargava uses his office primarily for meetings, the mathematical toys decorating its surfaces are more than just a colorful backdrop. When he was a graduate student at Princeton, they helped him solve a 200-year-old problem in number theory.

If two numbers that are each the sum of two perfect squares are multiplied together, the resulting number will also be the sum of two perfect squares (Try it!). As a child, Bhargava read in one of his grandfather's Sanskrit manuscripts about a generalization of this fact, developed in the year

628 by the great Indian mathematician Brahmagupta: If two numbers that are each the sum of a perfect square and a given whole number times a perfect square are multiplied together, the product will again be the sum of a perfect square and that whole number times another perfect square. "When I saw this math in my grandfather's manuscript, I got very excited," Bhargava said.

There are many other such relationships, in which numbers that can be expressed in a particular form can be multiplied together to produce a number with another particular form (sometimes the same form and sometimes a different one). As a graduate student, Bhargava discovered that in 1801, the German mathematical giant Carl Friedrich Gauss came up with a complete description of these kinds of relationships if the numbers can be expressed in what are known as binary quadratic forms: expressions with two variables and only quadratic terms, such as x^2+y^2 (the sum of two squares), x^2+7y^2 or $3x^2+4xy+9y^2$. Multiply two such expressions together, and Gauss' "composition law" tells you which quadratic form you will end up with. The only trouble is that Gauss' law is a mathematical behemoth, which took him about 20 pages to describe.

Bhargava wondered whether there was a simple way to describe what was going on and whether there were analogous laws for expressions involving higher exponents. He has always been drawn, he said, to questions like this one—"problems that are easy to state, and when you hear them, you think they're somehow so fundamental that we have to know the answer."

The answer came to him late one night as he was pondering the problem in his room, which was strewn with Rubik's Cubes and related puzzles, including the Rubik's mini-cube, which has only four squares on each face. Bhargava—who used to be able to solve the Rubik's Cube in about a minute—realized that if he were to place numbers on each corner of the mini-cube and then cut the cube in half, the eight corner numbers could be combined in a natural way to produce a binary quadratic form.

There are three ways to cut a cube in half—making a front-back, left-right or top-bottom division—so the cube generated three quadratic forms. These three forms, Bhargava discovered, add up to zero—not with respect to normal addition, but with respect to Gauss' method for composing quadratic forms. Bhargava's cube-slicing method gave a new and elegant reformulation of Gauss' 20-page law.

Additionally, Bhargava realized that if he arranged numbers on a Rubik's Domino—a 2x3x3 puzzle—he could produce a composition law for cubic forms, ones whose exponents are three. Over the next few years, Bhargava discovered 12 more composition laws, which formed the core of his Ph.D.

thesis.[1] These laws are not just idle curiosities: They connect to a fundamental object in modern number theory called an ideal class group, which measures how many ways a number can be factored into primes in more complicated number systems than the whole numbers.

"His Ph.D. thesis was phenomenal," Gross said. "It was the first major contribution to Gauss' theory of composition of binary forms for 200 years."

MAGICIAN

Bhargava's doctoral research earned him a five-year Clay Postdoctoral Fellowship, awarded by the Clay Mathematics Institute in Providence, Rhode Island, to new Ph.D.s who show leadership potential in mathematics research. He used the fellowship to spend one additional year at Princeton and the neighboring Institute for Advanced Study and then moved to Harvard. Only two years into his fellowship, however, job offers started pouring in, and a bidding war soon erupted over the young mathematician. "It was a crazy time," Bhargava said. At 28, he accepted a position at Princeton, becoming the second-youngest full professor in the university's history.

Back at Princeton, Bhargava felt like a graduate student again and had to be reminded by his former professors that he should call them by their first names now. "That was a little weird," he said. Bhargava ordered some frictionless chairs for his office, and he and his graduate student friends would race down the corridors of Fine Hall in the evenings. "One time, another professor happened to be there in the evening, and he came out of his office," Bhargava said. "That was rather embarrassing."

Bhargava is glad to be at an institution where he has the opportunity to teach. As an undergraduate teaching assistant at Harvard, he won the Derek C. Bok Award for excellence in teaching three years running. He especially enjoys reaching out to students in the arts or humanities, some of whom may think of themselves as mathphobic. "Because I came to math through the arts, it has been a passion of mine to bring in people who think of themselves as more on the art side than the science side," he said. Over the years, Bhargava has taught classes on the mathematics of music, poetry and magic. "I think anyone is reachable if the material is presented in the right way," he said.

Carolyn Chen, who took Bhargava's freshman seminar on mathematics and magic, called the course "super chill." Bhargava started each class by performing a magic trick—something he loves to do—and then the students dissected its mathematical principles. Bhargava's colleagues had warned him to steer clear of proofs, he said, "but by the end of the course, everyone was coming up with proofs without realizing that's what they were doing."

The course inspired Chen and several classmates to take more proof-based mathematics classes. "I took number theory after that freshman seminar," she said. "I would never have thought of taking it if I hadn't taken his class, but I really enjoyed it."

At Princeton, Bhargava started developing an arsenal of techniques for understanding the "geometry of numbers," a field somewhat akin to his childhood orange counting that studies how many points on a lattice lie inside a given shape. If the shape is fairly round and compact, like a pyramid of oranges, the number of lattice points inside the shape corresponds approximately to the shape's volume. But if the shape has long tentacles, it may capture many more—or many fewer—lattice points than a round shape of the same volume. Bhargava developed a way to understand the number of lattice points that appear in such tentacles.

"He has applied this method to one problem after another in number theory and just knocked them off," Gross said. "It's a beautiful thing to watch."

While Bhargava's early work on composition laws was a solo flight, much of his subsequent research has been in collaboration with others, something he describes as "a joyous experience." Working with Bhargava can be intense: At times, said Xiaoheng Wang in 2014, when he was a postdoctoral researcher at Princeton, he and Bhargava began discussing a math problem, and the next thing he knew, seven hours had passed. Characteristically, Bhargava is quick to deflect the honor of winning the Fields Medal onto his collaborators. "It's as much theirs as mine," he said.

In recent years, Bhargava has collaborated with several mathematicians to study elliptic curves, a type of equation whose highest exponent is three. Elliptic curves are one of the central objects in number theory: They were crucial to the proof of Fermat's Last Theorem, for example, and also have applications in cryptography.

A fundamental problem is to understand when such an equation has solutions that are whole numbers or ratios of whole numbers (rational numbers). Mathematicians have long known that most elliptic curves have either one rational solution or infinitely many, but they couldn't figure out, even after decades of trying, how many elliptic curves fall into each category. More recently, Bhargava has started to clear up this mystery. With Arul Shankar, his former doctoral student, Bhargava has shown that more than 20 percent of elliptic curves have exactly one rational solution.[2] And with Christopher Skinner, a colleague at Princeton, and Wei Zhang of

Columbia University, Bhargava has shown that at least 20 percent of elliptic curves have an infinite set of rational solutions with a particular structure called "rank 1."[3]

Bhargava, Skinner and Zhang have also made progress toward proving the famous Birch and Swinnerton-Dyer conjecture, a related problem about elliptic curves for which the Clay Mathematics Institute has offered a million-dollar prize. Bhargava, Skinner and Zhang have shown that the conjecture is true for more than 66 percent of elliptic curves.[4]

Bhargava's work on elliptic curves "has opened a whole world," Gross said. "Now everybody is excited about it and jumping in to work on it with him."

"He has proven some of the most exciting theorems in the past 20 years of number theory," Ono said. "The questions he attacks sound like things he shouldn't have the right to answer."

Bhargava has developed a unique mathematical style, Gross said. "You could look at a paper and say, 'Manjul's the only one who could have done that.' It's the mark of a really great mathematician that he doesn't have to sign his work."

THE ORACLE OF ARITHMETIC

Erica Klarreich

I n 2010, a startling rumor filtered through the number theory community and reached Jared Weinstein. Apparently, some graduate student at the University of Bonn in Germany had written a paper that redid "Harris–Taylor"—a 288-page book dedicated to a single impenetrable proof in number theory—in only 37 pages.[1] The 22-year-old student, Peter Scholze, had found a way to sidestep one of the most complicated parts of the proof, which deals with a sweeping connection between number theory and geometry.

"It was just so stunning for someone so young to have done something so revolutionary," said Weinstein, a number theorist now at Boston University. "It was extremely humbling."

Mathematicians at the University of Bonn, who made Scholze a full professor just two years later, were already aware of his extraordinary mathematical mind. After he posted his Harris–Taylor paper, experts in number theory and geometry started to notice Scholze too.

Since that time, Scholze, now 30, has risen to eminence in the broader mathematics community. Prize citations have called him "already one of the most influential mathematicians in the world" and "a rare talent which only emerges every few decades." In August 2018, he was awarded a Fields Medal, the highest honor in mathematics.

Scholze's key innovation—a class of fractal structures he calls perfectoid spaces—is only a few years old, but it already has far-reaching ramifications in the field of arithmetic geometry, where number theory and geometry come together. Scholze's work has a prescient quality, Weinstein said. "He can see the developments before they even begin."

Many mathematicians react to Scholze with "a mixture of awe and fear and exhilaration," said Bhargav Bhatt, a mathematician at the University of Michigan who has written joint papers with Scholze.

That's not because of his personality, which colleagues uniformly describe as grounded and generous. "He never makes you feel that he's, well, somehow so far above you," said Eugen Hellmann, Scholze's colleague at the University of Bonn.

Instead, it's because of his unnerving ability to see deep into the nature of mathematical phenomena. Unlike many mathematicians, he often starts not with a particular problem he wants to solve, but with some elusive concept that he wants to understand for its own sake. But then, said Ana Caraiani, a number theorist who has collaborated with Scholze, the structures he creates "turn out to have applications in a million other directions that weren't predicted at the time, just because they were the right objects to think about."

LEARNING ARITHMETIC

Scholze started teaching himself college-level mathematics at the age of 14, while attending Heinrich Hertz Gymnasium, a Berlin high school specializing in mathematics and science. At Heinrich Hertz, Scholze said, "you were not being an outsider if you were interested in mathematics."

At 16, Scholze learned that a decade earlier Andrew Wiles had proved the famous 17th-century problem known as Fermat's Last Theorem, which says that the equation $x^n + y^n = z^n$ has no nonzero whole-number solutions if n is greater than two. Scholze was eager to study the proof, but quickly discovered that despite the problem's simplicity, its solution uses some of the most cutting-edge mathematics around. "I understood nothing, but it was really fascinating," he said.

So Scholze worked backward, figuring out what he needed to learn to make sense of the proof. "To this day, that's to a large extent how I learn," he said. "I never really learned the basic things like linear algebra, actually—I only assimilated it through learning some other stuff."

As Scholze burrowed into the proof, he became captivated by the mathematical objects involved—structures called modular forms and elliptic curves that mysteriously unify disparate areas of number theory, algebra, geometry and analysis. Reading about the kinds of objects involved was perhaps even more fascinating than the problem itself, he said.

Scholze's mathematical tastes were taking shape. Today, he still gravitates toward problems that have their roots in basic equations about whole numbers. Those very tangible roots make even esoteric mathematical structures feel concrete to him. "I'm interested in arithmetic, in the end," he said. He's happiest, he said, when his abstract constructions lead him back around to small discoveries about ordinary whole numbers.

After high school, Scholze continued to pursue this interest in number theory and geometry at the University of Bonn. In his mathematics classes there, he never took notes, recalled Hellmann, who was his classmate.

Scholze could understand the course material in real time, Hellmann said. "Not just understand, but really understand on some kind of deep level, so that he also would not forget."

Scholze began doing research in the field of arithmetic geometry, which uses geometric tools to understand whole-number solutions to polynomial equations—equations such as $xy^2+3y=5$ that involve only numbers, variables and exponents. For some equations of this type, it is fruitful to study whether they have solutions among alternative number systems called p-adic numbers, which, like the real numbers, are built by filling in the gaps between whole numbers and fractions. But these systems are based on a nonstandard notion of where the gaps lie and which numbers are close to each other: In a p-adic number system, two numbers are considered close not if the difference between them is small, but if that difference is divisible many times by p.

It's a strange criterion, but a useful one. The 3-adic numbers, for example, provide a natural way to study equations like $x^2=3y^2$, in which factors of three are key.

P-adic numbers are "far removed from our everyday intuitions," Scholze said. Over the years, though, they have come to feel natural to him. "Now I find real numbers much, much more confusing than p-adic numbers. I've gotten so used to them that now real numbers feel very strange."

Mathematicians had noticed in the 1970s that many problems concerning p-adic numbers become easier if you expand the p-adic numbers by creating an infinite tower of number systems in which each one wraps around the one below it p times, with the p-adic numbers at the bottom of the tower. At the "top" of this infinite tower is the ultimate wraparound space—a fractal object that is the simplest example of the perfectoid spaces Scholze would later develop.

Scholze set himself the task of sorting out why this infinite wraparound construction makes so many problems about p-adic numbers and polynomials easier. "I was trying to understand the core of this phenomenon," he said. "There was no general formalism that could explain it."

He eventually realized that it's possible to construct perfectoid spaces for a wide variety of mathematical structures. These perfectoid spaces, he showed, make it possible to slide questions about polynomials from the p-adic world into a different mathematical universe in which arithmetic is much simpler (for instance, you don't have to carry when performing addition). "The weirdest property about perfectoid spaces is that they can magically move between the two number systems," Weinstein said.

This insight allowed Scholze to prove part of a complicated statement about the p-adic solutions to polynomials, called the weight-monodromy conjecture, which became his 2012 doctoral thesis.[2] The thesis "had such far-reaching implications that it was the topic of study groups all over the world," Weinstein said.

Scholze "found precisely the correct and cleanest way to incorporate all the previously done work and find an elegant formulation for that—and then, because he found really the correct framework, go way beyond the known results," Hellmann said.

Despite the complexity of perfectoid spaces, Scholze is known for the clarity of his talks and papers. "I don't really understand anything until Peter explains it to me," Weinstein said.

Scholze makes a point of trying to explain his ideas at a level that even beginning graduate students can follow, Caraiani said. "There's this sense of openness and generosity in terms of ideas," she said. "And he doesn't just do that with a few senior people, but really, a lot of young people have access to him." Scholze's friendly, approachable demeanor makes him an ideal leader in his field, Caraiani said. One time, when she and Scholze were on a difficult hike with a group of mathematicians, "he was the one running around making sure that everyone made it and checking up on everyone," Caraiani said.

Yet even with the benefit of Scholze's explanations, perfectoid spaces are hard for other researchers to grasp, Hellmann said. "If you move a little bit away from the path, or the way that he prescribes, then you're in the middle of the jungle and it's actually very hard." But Scholze himself, Hellmann said, "would never lose himself in the jungle, because he's never trying to fight the jungle. He's always looking for the overview, for some kind of clear concept."

Scholze avoids getting tangled in the jungle vines by forcing himself to fly above them: As when he was in college, he prefers to work without writing anything down. That means that he must formulate his ideas in the cleanest way possible, he said. "You have only some kind of limited capacity in your head, so you can't do too complicated things."

While other mathematicians are now starting to grapple with perfectoid spaces, some of the most far-reaching discoveries about them, not surprisingly, have come from Scholze and his collaborators. In 2013, a result he posted online "really kind of stunned the community," Weinstein said. "We had no idea that such a theorem was on the horizon."[3]

Scholze's result expanded the scope of rules known as reciprocity laws, which govern the behavior of polynomials that use the arithmetic of a clock (though not necessarily one with 12 hours). Clock arithmetics (in which, for example, $8 + 5 = 1$ if the clock has 12 hours) are the most natural and widely studied finite number systems in mathematics.

Reciprocity laws are generalizations of the 200-year-old quadratic reciprocity law, a cornerstone of number theory and one of Scholze's personal favorite theorems. The law states that given two prime numbers p and q, in most cases p is a perfect square on a clock with q hours exactly when q is a perfect square on a clock with p hours. For example, five is a perfect square on a clock with 11 hours, since $5 = 16 = 4^2$, and 11 is a perfect square on a clock with five hours, since $11 = 1 = 1^2$.

"I find it very surprising," Scholze said. "On the face of it, these two things seem to have nothing to do with each other."

"You can interpret a lot of modern algebraic number theory as just attempts to generalize this law," Weinstein said.

In the middle of the 20th century, mathematicians discovered an astonishing link between reciprocity laws and what seemed like an entirely different subject: the "hyperbolic" geometry of patterns such as M. C. Escher's famous angel-devil tilings of a disk. This link is a core part of the "Langlands program," a collection of interconnected conjectures and theorems about the relationship between number theory, geometry and analysis. When these conjectures can be proved, they are often enormously powerful: For instance, the proof of Fermat's Last Theorem boiled down to solving one small (but highly nontrivial) section of the Langlands program.

Mathematicians have gradually become aware that the Langlands program extends far beyond the hyperbolic disk; it can also be studied in higher-dimensional hyperbolic spaces and a variety of other contexts. Scholze has shown how to extend the Langlands program to a wide range of structures in "hyperbolic three-space"—a three-dimensional analogue of the hyperbolic disk—and beyond. By constructing a perfectoid version of hyperbolic three-space, Scholze has discovered an entirely new suite of reciprocity laws.

"Peter's work has really completely transformed what can be done, what we have access to," Caraiani said.

Scholze's result, Weinstein said, shows that the Langlands program is "deeper than we thought … it's more systematic, it's ever-present."

FAST FORWARD

Discussing mathematics with Scholze is like consulting a "truth oracle," according to Weinstein. "If he says, 'Yes, it is going to work,' you can be confident of it; if he says no, you should give right up; and if he says he doesn't know—which does happen—then, well, lucky you, because you've got an interesting problem on your hands."

Yet collaborating with Scholze is not as intense an experience as might be expected, Caraiani said. When she worked with Scholze, there was never a sense of hurry, she said. "It felt like somehow we were always doing things the right way—somehow proving the most general theorem that we could, in the nicest way, doing the right constructions that will illuminate things."

There was one occasion, though, when Scholze himself did hurry—while trying to finish up a paper in late 2013, shortly before the birth of his daughter. It was a good thing he pushed himself then, he said. "I didn't get much done afterwards."

Becoming a father has forced him to become more disciplined in how he uses his time, Scholze said. But he doesn't have to make a point of blocking off time for research—mathematics simply fills all the spaces between his other obligations. "Mathematics is my passion, I guess," he said. "I always want to think about it."

Yet he is not at all inclined to romanticize this passion. Asked if he felt he was meant to be a mathematician, he demurred. "That sounds too philosophical," he said.

A private person, he is somewhat uncomfortable with his growing celebrity (in March 2016, for example, he became the youngest recipient ever of Germany's prestigious Leibniz Prize, which awards 2.5 million euros to be used for future research). "At times it's a bit overwhelming," he said. "I try to not let my daily life get influenced by it."

Scholze continues to explore perfectoid spaces, but he has also branched out into other areas of mathematics touching on algebraic topology, which uses algebra to study shapes. Over the course of a year and a half, he became "a complete master of the subject," Bhatt said. "He changed the way [the experts] think about it."

It can be scary but also exciting for other mathematicians when Scholze enters their field, Bhatt said. "It means the subject is really going to move fast. I'm ecstatic that he's working in an area that's close to mine, so I actually see the frontiers of knowledge moving forward."

Yet to Scholze, his work thus far is just a warm-up. "I'm still in the phase where I'm trying to learn what's there, and maybe rephrasing it in my own words," he said. "I don't feel like I've actually started doing research."

AFTER PRIME PROOF, AN UNLIKELY STAR RISES

Thomas Lin

A s a boy in Shanghai, China, Yitang Zhang believed he would someday solve a great problem in mathematics. In 1964, at around the
age of 9, he found a proof of the Pythagorean theorem, which describes
the relationship between the lengths of the sides of any right triangle. He
was 10 when he first learned about two famous number theory problems,
Fermat's last theorem and the Goldbach conjecture. While he was not yet
aware of the centuries-old twin primes conjecture, he was already taken
with prime numbers, often described as indivisible "atoms" that make up
all other natural numbers.

But soon after, the anti-intellectual Cultural Revolution shuttered schools
and sent him and his mother to the countryside to work in the fields. Because
of his father's troubles with the Communist Party, Zhang was also unable to
attend high school. For 10 years, he worked as a laborer, reading books on
math, history and other subjects when he could.

Not long after the revolution ended, Zhang, then 23, enrolled at Peking
University and became one of China's top math students. After completing
his master's at the age of 29, he was recruited by T. T. Moh to pursue a doctorate at Purdue University in Lafayette, Indiana. But, promising though he
was, after defending his dissertation in 1991 he could not find academic
work as a mathematician.

In George Csicsery's documentary film *Counting From Infinity*, Zhang dis-
cusses his difficulties at Purdue and in the years that followed. He says his
doctoral adviser never wrote recommendation letters for him. (Moh has
written that Zhang did not ask for any.) Zhang admits that his shy, quiet
demeanor didn't help in building relationships or making himself known
to the wider math community. During this initial job-hunting period,
Zhang sometimes lived in his car, according to his friend Jacob Chi, music
director of the Pueblo Symphony in Colorado. In 1992, Zhang began working at another friend's Subway sandwich restaurant. For about seven years
he worked odd jobs for various friends.

In 1999, at 44, Zhang caught a break. A mathematician friend helped him secure work as a math lecturer at the University of New Hampshire. When he wasn't teaching his popular calculus classes, where students called him "Tom," he thought about number theory. By 2009, he had turned his attention to the twin primes conjecture, which postulates that there are an infinite number of prime number pairs with a difference of 2. Examples of twin prime pairs include 5 and 7, 11 and 13, and 17 and 19, but no one could prove that these pairs continue to exist all the way up the number line. In fact, no one could prove that there is any bounded prime gap at all, that primes don't just grow infinitely far apart.

In 2013, when the then-58-year-old Zhang published his proof of a bounded prime gap in the *Annals of Mathematics*, the paper's referees confirmed that he had proved "a landmark theorem in the distribution of prime numbers."

In the years since, Zhang has traveled the world giving talks and has received the Ostrowski Prize, the Cole Prize, the Rolf Schock Prize, a MacArthur fellowship and the attention of many major media outlets. Zhang fielded numerous job offers and eventually left the University of New Hampshire to become a professor of mathematics at the University of California, Santa Barbara. In February 2015, *Quanta Magazine* caught up with Zhang at the American Association for the Advancement of Science meeting in San Jose, California, where he gave a talk about advances in bounded prime gaps. The following interview was edited and condensed for clarity.

When and how did you first become aware that you were good at math?
When I was maybe the age of 9, maybe a little earlier, I was very interested in mathematics. I found the proof of Pythagoras's theorem. No one told me anything about that.

You were growing up in China—Shanghai—and later you weren't able to go to middle or high school.
Correct—because of the Cultural Revolution. At that time most of the people forgot about the science, the education. And instead, I was in the countryside just for the farm work. The revolution ended when I was 21. I went to Peking University when I was 23.

When you were not in school, how did you keep learning mathematics? Did you read books?
I read books. Actually, at that time I was also interested in lots of things. Not only math! Just reading every book I could get, like history and other topics.

Your background differs from that of most successful mathematicians. Even after you came to the U.S. and earned your doctorate, things didn't go so smoothly. For many years, you were doing accounting work, working for friends and not part of an academic setting.
Correct.

The math establishment didn't realize that, "OK, this is somebody we should nurture and cultivate"?
This is correct. I was not lucky.

What can be done to better identify people like you?
Maybe it is more important for a person to make himself known to the public. But that was not so easy for me. My personality didn't allow me to be very public, to be known by everyone, because maybe I'm too quiet.

There are other shy mathematicians who still seem to get the support they need.
These days, maybe it's easier. Historically, Riemann, Abel and many other famous mathematicians did not have such easy lives. They were not lucky.

What is it about the problem of prime gaps and the distribution of primes that is so interesting to you?
Problems like this are so interesting to every mathematician, I think, because we try to answer the essential problems of the mystery of numbers.

When you decide which problem to tackle, what are the criteria? Does it have to have a certain level of difficulty?
Yes, a certain level of difficulty. And an importance to mathematics. It's not that I say this is important, but that it is recognized as important by the mathematical community as a whole.

What is your approach to math beyond what you've said in other interviews—being patient and focused?
Do not easily say, "Oh, I really understand everything, so I have no problem." You try to discover problems, to ask yourself the problems. Then you can find a correct direction to solve the problem.

Keep asking questions? And keep an open mind?
Yes. An open mind.

What questions are you asking right now?
Still in the field of number theory, I may not have only one problem to think about, but a couple of problems, like the distribution of the zeros of the zeta functions and the L-functions.

Are you still thinking about the twin primes conjecture—getting the gap down to two?
That's not an easy problem. I didn't find a certain way to do it.

What would get the public more interested in mathematics?
Many problems—in number theory in particular—are easy for the public to understand. Even with some of the deeper mathematical problems, it is not difficult to understand the problem itself. That could help people to become more interested in mathematics.

When you picture a mathematician, you're probably not thinking of someone who's onstage and getting awards. What is your image of a mathematician?
Intuition. Your feelings in math. What is that? It's difficult to tell other people. That's your personal stuff.

Some of the big awards in math, particularly the Fields Medal, are aimed at younger mathematicians. You were in your mid-50s when you worked on bounded prime gaps.
I don't care so much about the age problem. I don't think there is a big difference. I can still do whatever I like to do.

When you were younger and first starting to get interested in math, did you ever imagine that you would solve a major problem like this?
Yes. When I was very young, I imagined there would be a day that I would solve a major math problem. I'm self-confident.

So you weren't necessarily surprised that you were able to solve the bounded prime gaps problem.
What surprised me was that my paper was recognized within three weeks. I hadn't expected that.

You were very busy afterward, traveling to universities and responding to media requests. Are you looking forward to a period with fewer talks and interviews—just focusing on the next problem?

I'm tired! I wish I could save my time and not spend too much of it being a star.

What do you hope to achieve over the next couple of decades?
I hope I can solve a few more important problems just like this.

IN NOISY EQUATIONS, ONE WHO HEARD MUSIC

Natalie Wolchover

M artin Hairer's masterwork is so fantastic, so fully baked and so far out of left field, one fellow mathematician declared, that the manuscript must have been downloaded into his brain by a more intelligent alien race.

Another compared the 180-page treatise to the *Lord of the Rings* trilogy because it "created a whole world." Few could recall another time in modern history when such a magnificent theory had emerged predominantly from the mind of a single person.

As for the researchers who have striven for decades to fathom the strange equations addressed by his theory, "He's taken them all to the cleaners," said Terry Lyons, a mathematician at the University of Oxford in England.

The Tolkienesque paper, published online in the journal *Inventiones Mathematicae* in March 2014, was only the latest and greatest in a series of feats by the then 38-year-old Hairer, who has frequently stunned colleagues with the speed and creativity of his work.[1] But if you were to take a seat next to Hairer at the pub near his home in Kenilworth, England, you might have a nice chat with him without ever suspecting the gangly, genial Austrian is one of the world's most brilliant mathematicians.

"Martin likes to talk to people; people like to talk to him," said his wife, the mathematician Xue-Mei Li. He is good-natured, knowledgeable and calm, she said—"and funny enough."

In August 2014, Hairer won a Fields Medal, widely viewed as the highest honor a mathematician can receive. Hairer, who was then a professor at the University of Warwick in England (and is now at Imperial College London), has been regarded since his late 20s as a leading figure in stochastic analysis, the branch of mathematics dealing with random processes like crystal growth and the spread of water in a napkin. Hairer's colleagues particularly note his rare mathematical intuition, an ability to sense the way toward grand solutions and beautiful proofs.

"If you leave him alone for a couple of days, he comes back with a miracle," said Hendrik Weber, a former colleague and collaborator at Warwick.

But, friends and colleagues say, this miracle worker's talents coexist with a disarmingly down-to-earth nature, extracurricular interests and even an entire career outside of mathematics. A lover of rock music and computer programming, Hairer is the creator of an award-winning sound-editing program called Amadeus, a popular tool among DJs, music producers and gaming companies and a lucrative sideline for Hairer.

"I don't think he has any of the stereotypes that the general public would like to assign to a mathematician," said Ofer Zeitouni, a professor of mathematics at the Weizmann Institute of Science in Israel who introduced Hairer's work at the 2014 International Congress of Mathematicians in Seoul, South Korea.

Hairer's well-roundedness has proven beneficial in a field that can seem detached from reality. It was his knowledge of a signal compression technique used in audio and image processing that inspired his otherworldly new theory.

The theory provides both the tools and the instruction manual for solving a huge class of previously unfathomable equations, statements that amount to "basically, 'infinity equals infinity,'" in one specialist's words, but which, despite their seeming senselessness, arise frequently in physics. The equations are mathematical abstractions of growth, the hustle and bustle of elementary particles and other "stochastic" processes, which evolve amid environmental noise.

It was these stochastic partial differential equations (SPDEs) that lured Hairer away from the path to a career as a physicist.

"It was just intriguing to me that you could derive these equations that don't make sense," he said in March 2014 during a stay at the Institute for Advanced Study in Princeton, New Jersey.

The mathematical inscrutability of many SPDEs has tantalized for decades. Their variables zigzag wildly through space and time, creating a mathematician's nightmare, a corner, at every point; worse still, to solve the equations, the infinite sharpness of those corners must somehow be multiplied and otherwise manipulated. In some cases, physicists have found tricks for approximating the solutions to the equations, such as ignoring the infinite jaggedness of the curves below a certain scale. But mathematicians have long sought a more rigorous understanding.

"The thing I've been working on is to give meaning to the equations," Hairer explained.

Hairer's theory of "regularity structures" brings order to SPDEs by broadening many of math's most basic concepts: derivatives, expansions and even what it means to be a solution. It is "kind of an extension of the classical calculus to this new setting," said Lorenzo Zambotti, a professor of mathematics at Pierre and Marie Curie University in Paris. Zambotti made the *Lord of the Rings* comparison upon reading a preliminary version of Hairer's paper in early 2013, and he has been studying the work ever since.

Experts describe Hairer's paper as at once clear and dense, a tightly woven exposition that other mathematicians will need time to unravel but which will probably be used for decades or centuries to come.

"Everyone knows it's brilliant," said David Kelly, a mathematician at New York University and a former Ph.D. student of Hairer's, "but everyone's also quite scared of it."

A LOGICAL MIND

A study in contrasts, Hairer is tall—6 feet 4 inches—but unassuming, with an angular face framing timid doe-eyes and a composure that is broken by frequent eruptions of hearty, boyish laughter. Hairer's list of accomplishments—especially the latest entry—may be intimidating, but he is not. "He's one of the least arrogant people I have ever met in my life," Weber said.

Although he considers mathematics his primary interest, Hairer "likes to shut it off" and says he gets many of his best ideas while contemplating other things. Even while scrawling equations on dry erase boards and zooming in and out of infinitely jagged lines on his laptop screen to explain his work, he switches with ease to casual, friendly conversation.

Outside of math, he enjoys reading Stephen King thrillers and other "silly books," cooking Asian-Western fusion dishes, skiing and going on frequent rambles through the countryside with Li. The two live in a semi-detached, Victorian house in Kenilworth and sometimes ride their bikes together the few miles to and from the Warwick campus.

According to Li, it is Hairer's diverse interests that inform his searing mathematical intuition, while his programming skills enable him to quickly test new ideas with algorithms. In her opinion, even his calm manner contributes to his success. "For large projects, people get overwhelmed, but he doesn't," she said. "He's good."

Behind Hairer's "normal personality," as many acquaintances put it, lies an uncommonly logical and organized mind. "Everything he learns he stores in an extremely structured way," Kelly said. As a graduate student, Kelly often stopped by Hairer's office to ask questions about stochastic analysis. "He would sort of stare off into the distance for 10 seconds and think about the question," Kelly said, "and then grab a piece of paper and deliver a textbook-standard answer in response—three pages of extremely detailed notes."

Hairer learned that he had won the Fields Medal during a visit to Columbia University in February 2014. "It's a big responsibility," he said a month later. "You sort of become an ambassador of mathematics."

Hairer said he did not expect to win and doesn't see himself as a typical Fields medalist. For starters, by his own reckoning, he was not a child prodigy, although he was clearly a very smart kid. Born into an Austrian family living in Switzerland, Hairer spent most of his childhood in Geneva, where his father, Ernst Hairer, works as a mathematician at the University of Geneva. Martin Hairer read chapter books by the precocious age of 6; became fluent in German, French and English; and performed at the top of his class all through school. "He was interested in everything," Ernst Hairer recalled.

For his 12th birthday, in 1987, Martin's father bought him a pocket calculator that could execute simple, 26-variable programs. He was instantly hooked. The next year, he persuaded his younger brother and sister to go in with him on a joint birthday present: a Macintosh II. He quickly became a proficient programmer, creating visualizations of fractals like the Mandelbrot set and then, at age 14, developing a program for solving ordinary differential equations—the much simpler cousins of SPDEs.

With his program, Hairer advanced to the national level of the European Union Contest for Young Scientists. The next year, he won a prize at the highest—the European—level of the contest with an interface for designing and simulating electrical circuits. At 16, in his final year of eligibility for the contest, he was interested in the physics of sound as well as Pink Floyd and The Beatles. He enjoyed recording musical notes and looking at the resulting waveforms on his computer and tried to write a program that could extract the notes from the recordings. The task was too difficult, but he ended up with a program for manipulating the recordings: version 1 of Amadeus. The software was selected for the European level of the competition, but the judges did not allow Hairer to advance a second time.

Pulled in different directions by mathematics, physics and computer science, Hairer only settled on math in his early 20s. At the time, he was working on a research project involving SPDEs as a Ph.D. student in physics at the University of Geneva. The mathematical aspects of the equations struck him as far more compelling than the physical phenomena they described. The equations seemed to possess a hidden meaning. After all, physicists had many black magic tricks for making their calculations work, which seemed to miraculously transform the equations into surprisingly close models of reality. But mathematically, they were ill-defined.

"Physicists are very good at being able to extract actual information from equations without bothering about whether they actually have a meaning,"

Hairer said, laughing. "They usually get it right, which is an amazing thing. But mathematicians like to actually know what the objects are."

He also relished the idea that if he succeeded in developing a mathematical theory of SPDEs, his discovery would hold forever.

"One advantage of mathematics is the immortality," he said. "Theorems that were proven 2,000 years ago are still true, whereas the physical worldview from 2,000 years ago definitely isn't."

Consider, for example, the divergent fates of Euclidean geometry—an ancient but enduring mathematical description of the architecture of flat space—and Aristotle's celestial spheres—imaginary concentric shells centered on the Earth that were thought to physically rotate the stars and planets through the heavens. "In physics, I could probably defend the reasoning behind a theory," Hairer said, "but I wouldn't defend it with my life."

REGULARITY IN RANDOMNESS

Hairer and Li met at a conference in 2001, at Warwick University, while he was a graduate student at Geneva and she was a professor at the University of Connecticut. "I enjoyed talking to him from the very beginning," said Li, who is originally from China. "I like the way he thinks and talks. Maybe I'm biased." After a few years of academic shuffling, the couple settled in Warwick, a collaborative and lively academic environment that suits them both.

As Hairer's career progressed, his talents became readily apparent, and according to experts in stochastic analysis, he has been "universally" renowned in the field for a decade.

His first major discovery came in 2004. Several groups were competing to prove that the two-dimensional stochastic Navier–Stokes equations—SPDEs that describe fluid flow in the presence of noise—are "ergodic," or eventually evolve to the same average state independent of their initial inputs. While riding the train on his way to meet with his collaborator on the project, Jonathan Mattingly of Duke University, Hairer had a sudden insight that grew into a powerful, case-closing result.[2]

In 2011, he solved a famous SPDE called the Kardar–Parisi–Zhang equation—a model of interface growth, such as the advancing edge of a bacterial colony in a petri dish and the spread of water in a napkin. The KPZ equation had been an open and much-studied problem since physicists proposed it in 1986.[3] Using an approach developed by Lyons called rough path theory, Hairer developed the core of his solution to the equation in less than a fortnight. Weber recalled that he went on vacation for 10 days right as

Hairer began working. "I came back, and he had solved the whole thing and had already typed it up," Weber said. "It was completely incredible to me."

Hairer's KPZ proof made a big splash, but by the time it appeared in the *Annals of Mathematics* in summer 2012, he was already in the thick of developing a more sophisticated approach for solving the KPZ equation, as well as even more complex SPDEs: his magnum opus, the theory of regularity structures.[4]

The problem with SPDEs is that they involve supremely thorny mathematical objects called "distributions." When a water droplet suffuses a napkin, for example, the advance of the water's edge depends on the current edge, as well as on noise: erratic factors like temperature variations and the creases and curves of the napkin. In the abstract form of an equation, the noise causes the edge to change infinitely quickly in space and time. And yet, according to the equation, the distribution that describes how quickly the edge changes in time is related to the square of the distribution describing how quickly it changes in space. But while smooth curves can easily be squared or divided, distributions do not submit to these arithmetic operations. "No object in the equations makes any sense classically," said Jeremy Quastel, a professor of mathematics at the University of Toronto.

For decades, mathematicians strove for a rigorous method of operating on distributions in order to solve SPDEs but made little headway. There are even published books that present incorrect procedures for doing so, Quastel said, "which is not something you would normally have in mathematics."

Hairer's big idea came to him in October 2011 while he was walking from the common room of the Warwick math department back to his office and thinking about nothing in particular. He suddenly realized that he could tame the distributions in SPDEs using an approach modeled on the mathematical properties of "wavelets"—brief, heartbeatlike oscillations that encode information in JPEG and MP3 files. Hairer had once considered using wavelets for a function in Amadeus. Conveniently for the purposes of data compression, any wavelet can be reconstructed by adding together a finite series of identical wavelets squashed to fractions of its initial width: a half, then a fourth, then an eighth and so on.

Similarly, Hairer realized, an infinitely jagged distribution like the ones that arise in SPDEs can also be written as a finite series. Each element of the series would consist of a set of curvelike objects that approximate the shape of the distribution at a fixed point in space and over a fixed time interval. In the next element in the series, this time interval would decrease to half, and then in the next to a fourth; as more elements in the series were included, the approximation would become more refined. Hairer suspected that, just as with wavelets, only a finite number of elements would be needed for the series to converge on the actual solution of the SPDE. If correct, he would be able to substitute the infinite and unfathomable distributions that arise in many SPDEs with a manageable number of perfectly calculable objects.

Right away, Hairer said, "I sort of knew it would work."

He went home and told Li about his epiphany. They got a textbook off the shelf and looked up wavelets, as neither of them knew the object's exact mathematical definition. Li soon saw that Hairer's idea was brilliant. "I said he should pursue it and spend lots of time on it," she said. "Instead of going out, just sit down and work."

STOCHASTIC FUTURE

By the time Andrew Wiles proved Fermat's Last Theorem in 1995, the 358-year-old problem had generated so much activity throughout its history that fame and recognition for Wiles were instantaneous. In Hairer's case, no one reasonably expected a general theory of SPDEs. It seemed to come out of nowhere. "I suppose I kind of created a flurry of activity," Hairer said.

Although dozens of mathematicians in Hairer's immediate research area are learning his theory, some experts fear that it is too technically challenging to gain widespread use. "There is a worry that it won't have the impact it deserves, not through any fault of Martin's but just because the simplest way that one can deal with this type of problem is just too difficult to be popularizable," said Quastel, who joked to colleagues that the theory must have been a dispatch from aliens. The power of regularity structures is easily understood, but when Hairer actually delves into how the theory works in talks and papers, Quastel said, "he loses his audience a little bit."

If the theory does take hold, Lyons said, a deeper understanding of SPDEs could someday become useful in real physical models, such as in particle physics and machine learning. "There are an enormous number of situations where you have complex spatial behavior with randomness, where there is physical or social significance to understanding what the hell's going on," he said, "and I think Martin has made revolutionary contributions to our ability to tackle those things mathematically."

The possibility that his theory might have physical relevance seems to hold little allure for Hairer, however. When asked whether he thought regularity structures might reveal something new about the "actual universe," he merely laughed heartily. He said he finds joy in the newly discernible properties of the equations themselves, such as how far the solutions fluctuate away from their average value over long stretches of time or space or how rapidly two solutions with different starting points wriggle toward each other.

Picture these squirming solutions plotted side by side on the same graph—mathematical abstractions of, say, what would happen if water dripped onto two identical napkins.

"At some point they touch," Hairer said, referring to such a pair of solutions. "You can ask how snugly do they fit each other at the point where they touch. And it turns out they are much snugger than you would think." He laughed with delight.

MICHAEL ATIYAH'S IMAGINATIVE STATE OF MIND

Siobhan Roberts

Despite Michael Atiyah's many accolades—he is a winner of both the Fields and the Abel prizes for mathematics; a past president of the Royal Society of London, the oldest scientific society in the world (and a past president of the Royal Society of Edinburgh); a former master of Trinity College, Cambridge; a knight and a member of the royal Order of Merit; and essentially Britain's mathematical pope—he is nonetheless perhaps most aptly described as a matchmaker. He has an intuition for arranging just the right intellectual liaisons, oftentimes involving himself and his own ideas, and over the course of his half-century-plus career he has bridged the gap between apparently disparate ideas within the field of mathematics, and between mathematics and physics.

One day in the spring of 2013, for instance, as he sat in the Queen's Gallery at Buckingham Palace awaiting the annual Order of Merit luncheon with Elizabeth II, Sir Michael made a match for his lifelong friend and colleague, Sir Roger Penrose, the great mathematical physicist.

Penrose had been trying to develop his "twistor" theory, a path toward quantum gravity that's been in the works for nearly 50 years. "I had a way of doing it which meant going out to infinity," Penrose said, "and trying to solve a problem out there, and then coming back again." He thought there must be a simpler way. And right then and there Atiyah put his finger on it, suggesting Penrose make use of a type of "noncommutative algebra."

"I thought, 'Oh, my God,'" Penrose said. "Because I knew there was this noncommutative algebra which had been sitting there all this time in twistor theory. But I hadn't thought of using it in this particular way. Some people might have just said, 'That won't work.' But Michael could immediately see that there was a way in which you could make it work, and exactly the right thing to do." Given the venue where Atiyah made the suggestion, Penrose dubbed his improved idea "palatial twistor theory."[1]

This is the power of Atiyah. Roughly speaking, he has spent the first half of his career connecting mathematics to mathematics, and the second half connecting mathematics to physics.

Atiyah is best known for the "index theorem," devised in 1963 with Isadore Singer of the Massachusetts Institute of Technology (and properly called the Atiyah–Singer index theorem), connecting analysis and topology—a fundamental connection that proved to be important in both mathematical fields, and later in physics as well.[2] Largely for this work, Atiyah won the Fields Medal in 1966 and the Abel Prize in 2004 (with Singer).

In the 1980s, methods gleaned from the index theorem unexpectedly played a role in the development of string theory—an attempt to reconcile the large-scale realm of general relativity and gravity with the small-scale realm of quantum mechanics—particularly with the work of Edward Witten, a string theorist at the Institute for Advanced Study in Princeton. Witten and Atiyah began an extended collaboration, and in 1990 Witten won the Fields Medal, the only physicist ever to win the prize, with Atiyah as his champion.

In late 2015 and early 2016, when he sat for interviews with *Quanta Magazine* at the age of 86, Atiyah was hardly lowering the bar. He was still tackling the big questions, still trying to orchestrate a union between the quantum and the gravitational forces. On this front, the ideas were arriving fast and furious, but as Atiyah himself described, they were as yet intuitive, imaginative, vague and clumsy commodities.

Still, he was relishing this state of free-flowing creativity, energized by his packed schedule. In hot pursuit of these lines of investigation and contemplation, in December 2015 he delivered a double-header of lectures, back-to-back on the same day, at the University of Edinburgh, where he has been an honorary professor since 1997. He was keen to share his new ideas and, he hopes, attract supporters. To that end, the month before he hosted a conference at the Royal Society of Edinburgh on "The Science of Beauty." *Quanta* interviewed Atiyah at the Royal Society gathering and afterward, whenever he slowed down long enough to take questions. What follows is an edited and condensed version of those catch-as-catch-can conversations.

Where do you trace the beginnings of your interest in beauty and science?
I was born 86 years ago. That's when my interest started. I was conceived in Florence. My parents were going to name me Michelangelo, but someone said, "That's a big name for a small boy." It would have been a disaster. I can't draw. I have no talent at all.

You mentioned that something "clicked" during Roger Penrose's lecture on "The Role of Art in Mathematics" and that you now have an idea for a collaborative paper. What is this clicking, the process or the state—can you describe it?

It's the kind of thing that once you've seen it, the truth or veracity, it just stares you in the face. The truth is looking back at you. You don't have to look for it. It's shining on the page.

Is that generally how your ideas arrive?

This was a spectacular version. The crazy part of mathematics is when an idea appears in your head. Usually when you're asleep, because that's when you have the fewest inhibitions. The idea floats in from heaven knows where. It floats around in the sky; you look at it, and admire its colors. It's just there. And then at some stage, when you try to freeze it, put it into a solid frame, or make it face reality, then it vanishes, it's gone. But it's been replaced by a structure, capturing certain aspects, but it's a clumsy interpretation.

Have you always had mathematical dreams?

I think so. Dreams happen during the daytime, they happen at night. You can call them a vision or intuition. But basically they're a state of mind—without words, pictures, formulas or statements. It's "pre" all that. It's pre-Plato. It's a very primordial feeling. And again, if you try to grasp it, it always dies. So when you wake up in the morning, some vague residue lingers, the ghost of an idea. You try to remember what it was and you only get half of it right, and maybe that's the best you can do.

Is imagination part of it?

Absolutely. Time travel in the imagination is cheap and easy—you don't even need to buy a ticket. People go back and imagine they are part of the Big Bang, and then they ask the question of what came before.

What guides the imagination—beauty?

It's not the kind of beauty that you can point to—it's beauty in a much more abstract sense.

Not too long ago you published a study, with Semir Zeki, a neurobiologist at University College London, and other collaborators, on "The Experience of Mathematical Beauty and Its Neural Correlates."[3]

That's the most-read article I've ever written! It's been known for a long time that some part of the brain lights up when you listen to nice music, or

read nice poetry, or look at nice pictures—and all of those reactions happen
in the same place [the "emotional brain," specifically the medial orbito-
frontal cortex]. And the question was: Is the appreciation of mathematical
beauty the same, or is it different? And the conclusion was, it is the same.
The same bit of the brain that appreciates beauty in music, art and poetry
is also involved in the appreciation of mathematical beauty. And that was
a big discovery.

**You reached this conclusion by showing mathematicians various equations
while a functional MRI recorded their response. Which equation won out
as most beautiful?**
Ah, the most famous of all, Euler's equation:

$$e^{i\pi} + 1 = 0$$

FIGURE 4.1

It involves π; the mathematical constant e [Euler's number, 2.71828...];
i, the imaginary unit; 1; and 0—it combines all the most important things
in mathematics in one formula, and that formula is really quite deep. So
everybody agreed that that was the most beautiful equation. I used to say it
was the mathematical equivalent of Hamlet's phrase "To be, or not to be"—
very short, very succinct, but at the same time very deep. Euler's equation
uses only five symbols, but it also encapsulates beautifully deep ideas, and
brevity is an important part of beauty.

**You are especially well-known for two supremely beautiful works, not only
the index theorem but also K-theory, developed with the German topologist Friedrich Hirzebruch. Tell me about K-theory.**
The index theorem and K-theory are actually two sides of the same coin. They
started out different, but after a while they became so fused together that you
can't disentangle them. They are both related to physics, but in different ways.
    K-theory is the study of flat space, and of flat space moving around. For
example, let's take a sphere, the Earth, and let's take a big book and put it
on the Earth and move it around. That's a flat piece of geometry moving
around on a curved piece of geometry. K-theory studies all aspects of that
situation—the topology and the geometry. It has its roots in our navigation
of the Earth.
    The maps we used to explore the Earth can also be used to explore
both the large-scale universe, going out into space with rockets, and the

small-scale universe, studying atoms and molecules. What I'm doing now is trying to unify all that, and K-theory is the natural way to do it. We've been doing this kind of mapping for hundreds of years, and we'll probably be doing it for thousands more.

Did it surprise you that K-theory and the index theorem turned out to be important in physics?
Oh, yes. I did all this geometry not having any notion that it would be linked to physics. It was a big surprise when people said, "Well, what you're doing is linked to physics." And so I learned physics quickly, talking to good physicists to find out what was happening.

How did your collaboration with Witten come about?
I met him in Boston in 1977, when I was getting interested in the connection between physics and mathematics. I attended a meeting, and there was this young chap with the older guys. We started talking, and after a few minutes I realized that the younger guy was much smarter than the old guys. He understood all the mathematics I was talking about, so I started paying attention to him. That was Witten. And I've kept in touch with him ever since.

What was he like to work with?
In 2001, he invited me to Caltech, where he was a visiting professor. I felt like a graduate student again. Every morning I would walk into the department, I'd go to see Witten, and we'd talk for an hour or so. He'd give me my homework. I'd go away and spend the next 23 hours trying to catch up. Meanwhile, he'd go off and do half a dozen other things. We had a very intense collaboration. It was an incredible experience because it was like working with a brilliant supervisor. I mean, he knew all the answers before I got them. If we ever argued, he was right and I was wrong. It was embarrassing!

You've said before that the unexpected interconnections that pop up occasionally between math and physics are what appeal to you most—you like finding yourself wading into unfamiliar territory.
Right; well, you see, a lot of mathematics is predictable. Somebody shows you how to solve one problem, and you do the same thing again. Every time you take a step forward you're following in the steps of the person who came before. Every now and again, somebody comes along with a totally new idea and shakes everybody up. To start with, people don't believe it, and then when they do believe it, it leads in a totally new direction. Mathematics comes in fits and starts. It has continuous development, and then it

has discontinuous jumps, when suddenly somebody has a new idea. Those are the ideas that really matter. When you get them, they have major consequences. We're about due another one. Einstein had a good idea 100 years ago, and we need another one to take us forward.[4]

But the approach has to be more investigative than directive. If you try to direct science, you only get people going in the direction you told them to go. All of science comes from people noticing interesting side paths. You've got to have a very flexible approach to exploration and allow different people to try different things. Which is difficult, because unless you jump on the bandwagon, you don't get a job.

Worrying about your future, you have to stay in line. That's the worst thing about modern science. Fortunately, when you get to my age, you don't need to bother about that. I can say what I like.

These days, you're trying out some new ideas in hopes of breaking the stalemate in physics?
Well, you see, there's atomic physics—electrons and protons and neutrons, all the stuff of which atoms are made. At these very, very, very small scales, the laws of physics are much the same, but there is also a force you ignore, which is the gravitational force. Gravity is present everywhere because it comes from the entire mass of the universe. It doesn't cancel itself out, it doesn't have positive or negative value, it all adds up. So however far away the black holes and galaxies are, they all exert a very small force everywhere in the universe, even in an electron or proton. But physicists say, "Ah, yes, but it's so small you can ignore it; we don't measure things that small, we do perfectly well without it." My starting point is that that is a mistake. If you correct that mistake, you get a theory that is much better.

I'm now looking again at some of the ideas that were around 100 years ago and that were discarded at the time because people couldn't understand what the ideas were trying to get at. How does matter interact with gravity? Einstein's theory was that if you put a bit of matter in, it changes the curvature of space. And when the curvature of space changes, it acts on the matter. It's a very complicated feedback mechanism.

I'm going back to Einstein and [Paul] Dirac and looking at them again with new eyes, and I think I'm seeing things that people missed. I'm filling in the holes of history, taking account of new discoveries. Archaeologists dig things up, or historians find a new manuscript, and that sheds an entirely new light. So that's what I've been doing. Not by going into libraries, but by sitting in my room at home, thinking. If you think long enough, you get a good idea.

So you're saying that the gravitational force can't be ignored?
I think all the difficulty physicists have had comes from ignoring that. You shouldn't ignore it. And the point is, I believe the mathematics gets simplified if you feed it in. If you leave it out, you make things more difficult for yourself.

Most people would say you don't need to worry about gravitation when you look at atomic physics. The scale is so small that, for the kind of calculations we do, it can be ignored. In some sense, if you just want answers, that's correct. But if you want understanding, then you've made a mistake in that choice.

If I'm wrong, well, I made a mistake. But I don't think so. Because once you pick this idea up, there are all sorts of nice consequences. The mathematics fits together. The physics fits together. The philosophy fits together.

What does Witten think of your new ideas?
Well, it's a challenge. Because when I talked to him in the past about some of my ideas, he dismissed them as hopeless, and he gave me 10 different reasons why they're hopeless. Now I think I can defend my ground. I've spent a lot of time thinking, coming at it from different angles and coming back to it. And I'm hoping I can persuade him that there is merit to my new approach.

You're risking your reputation, but you think it's worth it.
My reputation is established as a mathematician. If I make a mess of it now, people will say, "All right, he was a good mathematician, but at the end of his life he lost his marbles."

A friend of mine, John Polkinghorne, left physics just as I was going in; he went into the church and became a theologian. We had a discussion on my 80th birthday and he said to me, "You've got nothing to lose; you just go ahead and think what you think." And that's what I've been doing. I've got all the medals I need. What could I lose? So that's why I'm prepared to take a gamble that a young researcher wouldn't be prepared to take.

Are you surprised to be so charged up about new ideas at this stage of your career?
One of my sons said to me, "Impossible, Dad. Mathematicians do all their best work by the time they're 40. And you're over 80. It's impossible for you to have a good idea now."

If you're still awake and alert mentally when you're over 80, you've got the advantage that you've lived a long time and you've seen many things, and you get perspective. I'm 86 now, and it's in the last few years that I've had these ideas. New ideas come along and you pick up bits here and there, and the time is ripe now, whereas it might not have been ripe five or 10 years ago.

Is there one big question that has always guided you?
I always want to try to understand why things work. I'm not interested in getting a formula without knowing what it means. I always try to dig behind the scenes, so if I have a formula, I understand why it's there. And understanding is a very difficult notion.

People think mathematics begins when you write down a theorem followed by a proof. That's not the beginning, that's the end. For me the creative place in mathematics comes before you start to put things down on paper, before you try to write a formula. You picture various things, you turn them over in your mind. You're trying to create, just as a musician is trying to create music, or a poet. There are no rules laid down. You have to do it your own way. But at the end, just as a composer has to put it down on paper, you have to write things down. But the most important stage is understanding. A proof by itself doesn't give you understanding. You can have a long proof and no idea at the end of why it works. But to understand why it works, you have to have a kind of gut reaction to the thing. You've got to feel it.

V WHAT CAN OR CAN'T COMPUTERS DO?

HACKER-PROOF CODE CONFIRMED

Kevin Hartnett

I n the summer of 2015 a team of hackers attempted to take control of an unmanned military helicopter known as Little Bird. The helicopter, which is similar to the piloted version long-favored for U.S. special operations missions, was stationed at a Boeing facility in Arizona. The hackers had a head start: At the time they began the operation, they already had access to one part of the drone's computer system. From there, all they needed to do was hack into Little Bird's onboard flight-control computer, and the drone was theirs.

When the project started, a "Red Team" of hackers could have taken over the helicopter almost as easily as it could break into your home Wi-Fi. But in the intervening months, engineers from the Defense Advanced Research Projects Agency (DARPA) had implemented a new kind of security mechanism—a software system that couldn't be commandeered. Key parts of Little Bird's computer system were unhackable with existing technology, its code as trustworthy as a mathematical proof. Even though the Red Team was given six weeks with the drone and more access to its computing network than genuine bad actors could ever expect to attain, they failed to crack Little Bird's defenses.

"They were not able to break out and disrupt the operation in any way," said Kathleen Fisher, a professor of computer science at Tufts University and the founding program manager of the High-Assurance Cyber Military Systems (HACMS) project. "That result made all of DARPA stand up and say, oh my goodness, we can actually use this technology in systems we care about."

The technology that repelled the hackers was a style of software programming known as formal verification. Unlike most computer code, which is written informally and evaluated based mainly on whether it works, formally verified software reads like a mathematical proof: Each statement follows logically from the preceding one. An entire program can be tested with the same certainty that mathematicians prove theorems.

"You're writing down a mathematical formula that describes the program's behavior and using some sort of proof checker that's going to check the correctness of that statement," said Bryan Parno, who has done research on formal verification and security at Microsoft Research and at Carnegie Mellon University.

The aspiration to create formally verified software has existed nearly as long as the field of computer science. For a long time it seemed hopelessly out of reach, but advances over the past decade in so-called "formal methods" have inched the approach closer to mainstream practice. Today formal software verification is being explored in well-funded academic collaborations, the U.S. military and technology companies such as Microsoft and Amazon.

The interest occurs as an increasing number of vital social tasks are transacted online. Previously, when computers were isolated in homes and offices, programming bugs were merely inconvenient. Now those same small coding errors open massive security vulnerabilities on networked machines that allow anyone with the know-how free rein inside a computer system.

"Back in the 20th century, if a program had a bug, that was bad, the program might crash, so be it," said Andrew Appel, professor of computer science at Princeton University and a leader in the program verification field. But in the 21st century, a bug could create "an avenue for hackers to take control of the program and steal all your data. It's gone from being a bug that's bad but tolerable to a vulnerability, which is much worse," he said.

THE DREAM OF PERFECT PROGRAMS

In October 1973 Edsger Dijkstra came up with an idea for creating error-free code. While staying in a hotel at a conference, he found himself seized in the middle of the night by the idea of making programming more mathematical. As he explained in a later reflection, "With my brain burning, I left my bed at 2:30 a.m. and wrote for more than an hour." That material served as the starting point for his seminal 1976 book, *A Discipline of Programming*, which, together with work by Tony Hoare (who, like Dijkstra, received the Turing Award, computer science's highest honor), established a vision for incorporating proofs of correctness into how computer programs are written.

It's not a vision that computer science followed, largely because for many years afterward it seemed impractical—if not impossible—to specify a program's function using the rules of formal logic.

A formal specification is a way of defining what, exactly, a computer program does. And a formal verification is a way of proving beyond a doubt that a program's code perfectly achieves that specification. To see how this works, imagine writing a computer program for a robot car that drives you to the grocery store. At the operational level, you'd define the moves the car has at its disposal to achieve the trip—it can turn left or right, brake or accelerate, turn on or off at either end of the trip. Your program, as it were, would be a compilation of those basic operations arranged in the appropriate order so that at the end, you arrived at the grocery store and not the airport.

The traditional, simple way to see if a program works is to test it. Coders submit their programs to a wide range of inputs (or unit tests) to ensure they behave as designed. If your program were an algorithm that routed a robot car, for example, you might test it between many different sets of points. This testing approach produces software that works correctly, most of the time, which is all we really need for most applications. But unit testing can't guarantee that software will always work correctly because there's no way to run a program through every conceivable input. Even if your driving algorithm works for every destination you test it against, there's always the possibility that it will malfunction under some rare conditions—or "corner cases," as they're called—and open a security gap. In actual programs, these malfunctions could be as simple as a buffer overflow error, where a program copies a little more data than it should and overwrites a small piece of the computer's memory. It's a seemingly innocuous error that's hard to eliminate and provides an opening for hackers to attack a system—a weak hinge that becomes the gateway to the castle.

"One flaw anywhere in your software, and that's the security vulnerability. It's hard to test every possible path of every possible input," Parno said.

Actual specifications are subtler than a trip to the grocery store. Programmers may want to write a program that notarizes and time-stamps documents in the order in which they're received (a useful tool in, say, a patent office). In this case the specification would need to explain that the counter always increases (so that a document received later always has a higher number than a document received earlier) and that the program will never leak the key it uses to sign the documents.

This is easy enough to state in plain English. Translating the specification into formal language that a computer can apply is much harder—and accounts for a main challenge when writing any piece of software in this way.

"Coming up with a formal machine-readable specification or goal is conceptually tricky," Parno said. "It's easy to say at a high level 'don't leak my password,' but turning that into a mathematical definition takes some thinking."

To take another example, consider a program for sorting a list of numbers. A programmer trying to formalize a specification for a sort program might come up with something like this:

For every item j in a list, ensure that the element $j \leq j+1$

Yet this formal specification—ensure that every element in a list is less than or equal to the element that follows it—contains a bug: The programmer assumes that the output will be a permutation of the input. That is, given the list [7, 3, 5], she expects that the program will return [3, 5, 7] and satisfy the definition. Yet the list [1, 2] also satisfies the definition since "it is *a* sorted list, just not the sorted list we were probably hoping for," Parno said.

In other words, it's hard to translate an idea you have for what a program should do into a formal specification that forecloses every possible (but incorrect) interpretation of what you want the program to do. And the example above is for something as simple as a sort program. Now imagine taking something much more abstract than sorting, such as protecting a password. "What does that mean mathematically? Defining it may involve writing down a mathematical description of what it means to keep a secret, or what it means for an encryption algorithm to be secure," Parno said. "These are all questions we, and many others, have looked at and made progress on, but they can be quite subtle to get right."

BLOCK-BASED SECURITY

Between the lines it takes to write both the specification and the extra annotations needed to help the programming software reason about the code, a program that includes its formal verification information can be five times as long as a traditional program that was written to achieve the same end.

This burden can be alleviated somewhat with the right tools—programming languages and proof-assistant programs designed to help software engineers construct bombproof code. But those didn't exist in the 1970s. "There were many parts of science and technology that just weren't mature enough to make that work, and so around 1980, many parts of the computer science field lost interest in it," said Appel, who is the lead principal investigator of a research group called DeepSpec that's developing formally verified computer systems.

Even as the tools improved, another hurdle stood in the way of program verification: No one was sure whether it was even necessary. While formal methods enthusiasts talked of small coding errors manifesting as catastrophic bugs, everyone else looked around and saw computer pro-grams that pretty much worked fine. Sure, they crashed sometimes, but losing a little unsaved work or having to restart occasionally seemed like a small price to pay for not having to tediously spell out every little piece of a program in the language of a formal logical system. In time, even program verification's earliest champions began to doubt its usefulness. In the 1990s Hoare—whose "Hoare logic" was one of the first formal systems for rea-soning about the correctness of a computer program—acknowledged that maybe specification was a labor-intensive solution to a problem that didn't exist. As he wrote in 1995:

> Ten years ago, researchers into formal methods (and I was the most mistaken among them) predicted that the programming world would embrace with grati-tude every assistance promised by formalization....It has turned out that the world just does not suffer significantly from the kind of problem that our research was originally intended to solve.

Then came the internet, which did for coding errors what air travel did for the spread of infectious diseases: When every computer is connected to every other one, inconvenient but tolerable software bugs can lead to a cascade of security failures.

"Here's the thing we didn't quite fully understand," Appel said. "It's that there are certain kinds of software that are outward-facing to all hackers on the internet, so that if there is a bug in that software, it might well be a security vulnerability."

By the time researchers began to understand the critical threats to com-puter security posed by the internet, program verification was ready for a comeback. To start, researchers had made big advances in the technol-ogy that undergirds formal methods: improvements in proof-assistant programs like Coq and Isabelle that support formal methods; the develop-ment of new logical systems (called dependent-type theories) that provide a framework for computers to reason about code; and improvements in what's called "operational semantics"—in essence, a language that has the right words to express what a program is supposed to do.

"If you start with an English-language specification, you're inherently starting with an ambiguous specification," said Jeannette Wing, a former vice president at Microsoft Research and is now a professor of computer sci-ence at Columbia University. "Any natural language is inherently ambigu-ous. In a formal specification you're writing down a precise specification based on mathematics to explain what it is you want the program to do."

In addition, researchers in formal methods also moderated their goals. In the 1970s and early 1980s, they envisioned creating entire fully verified computer systems, from the circuit all the way to the programs. Today most formal methods researchers focus instead on verifying smaller but especially vulnerable or critical pieces of a system, like operating systems or cryptographic protocols.

"We're not claiming we're going to prove an entire system is correct, 100 percent reliable in every bit, down to the circuit level," Wing said. "That's ridiculous to make those claims. We are much more clear about what we can and cannot do."

The HACMS project illustrates how it's possible to generate big security guarantees by specifying one small part of a computer system. The project's first goal was to create an unhackable recreational quadcopter. The off-the-shelf software that ran the quadcopter was monolithic, meaning that if an attacker broke into one piece of it, he had access to all of it. So, over the next two years, the HACMS team set about dividing the code on the quadcopter's mission-control computer into partitions.

The team also rewrote the software architecture, using what Fisher, the HACMS founding program manager, calls "high-assurance building blocks"— tools that allow programmers to prove the fidelity of their code. One of those verified building blocks comes with a proof guaranteeing that someone with access inside one partition won't be able to escalate their privileges and get inside other partitions.

Later the HACMS programmers installed this partitioned software on Little Bird. In the test against the Red Team hackers, they provided the Red Team access inside a partition that controlled aspects of the drone helicopter, like the camera, but not essential functions. The hackers were mathematically guaranteed to get stuck. "They proved in a machine-checked way that the Red Team would not be able to break out of the partition, so it's not surprising" that they couldn't, Fisher said. "It's consistent with the theorem, but it's good to check."

Since the Little Bird test, DARPA has been applying the tools and techniques from the HACMS project to other areas of military technology, like satellites and self-driving convoy trucks. The new initiatives are consistent with the way formal verification has spread over the last decade: Each successful project emboldens the next. "People can't really have the excuse anymore that it's too hard," Fisher said.

VERIFYING THE INTERNET

Security and reliability are the two main goals that motivate formal methods. And with each passing day the need for improvements in both is more apparent. In 2014 a small coding error that would have been caught by formal specification opened the way for the Heartbleed bug, which threatened to bring down the internet. A year later a pair of white-hat hackers confirmed perhaps the biggest fears we have about internet-connected cars when they successfully took control of someone else's Jeep Cherokee.

As the stakes rise, researchers in formal methods are pushing into more ambitious places. In a return to the spirit that animated early verification efforts in the 1970s, the DeepSpec collaboration led by Appel (who also worked on HACMS) is attempting to build a fully verified end-to-end system like a web server. If successful, the effort, which is funded by a $10 million grant from the National Science Foundation, would stitch together many of the smaller-scale verification successes of the last decade. Researchers have built a number of provably secure components, such as the core, or kernel, of an operating system. "What hadn't been done, and is the challenge DeepSpec is focusing on, is how to connect those components together at specification interfaces," Appel said.

Over at Microsoft Research, software engineers have two ambitious formal verification projects under way. The first, named Everest, is to create a verified version of HTTPS, the protocol that secures web browsers and that Wing refers to as the "Achilles heel of the internet."

The second is to create verified specifications for complex cyber-physical systems such as drones. Here the challenge is considerable. Where typical software follows discrete, unambiguous steps, the programs that tell a drone how to move use machine learning to make probabilistic decisions based on a continuous stream of environmental data. It's far from obvious how to reason about that kind of uncertainty or pin it down in a formal specification. But formal methods have advanced a lot even in the last decade, and Wing, who oversees this work, is optimistic formal methods researchers are going to figure it out.

WILL COMPUTERS REDEFINE THE ROOTS OF MATH?

Kevin Hartnett

O n a train trip from Lyon to Paris in 2014, Vladimir Voevodsky sat next to Steve Awodey and tried to convince him to change the way he does mathematics.

Voevodsky, then 48, was a permanent faculty member at the Institute for Advanced Study. He was born in Moscow but spoke nearly flawless English, and he had the confident bearing of someone with no need to prove himself to anyone. In 2002 he won the Fields Medal, which is often considered the most prestigious award in mathematics.

As their train approached the city, Voevodsky pulled out his laptop and opened a program called Coq, a proof assistant that provides mathematicians with an environment in which to write mathematical arguments. Awodey, a mathematician and logician at Carnegie Mellon University in Pittsburgh, followed along as Voevodsky wrote a definition of a mathematical object using a new formalism he had created, called univalent foundations. It took Voevodsky 15 minutes to write the definition.

"I was trying to convince [Awodey] to do [his mathematics in Coq]," Voevodsky explained during a fall 2014 lecture. "I was trying to convince him that it's easy to do."

The idea of doing mathematics in a program like Coq has a long history. The appeal is simple: Rather than relying on fallible human beings to check proofs, you can turn the job over to computers, which can tell whether a proof is correct with complete certainty. Despite this advantage, computer proof assistants haven't been widely adopted in mainstream mathematics. This is partly because translating everyday math into terms a computer can understand is cumbersome and, in the eyes of many mathematicians, not worth the effort.

For nearly a decade, Voevodsky had been advocating the virtues of computer proof assistants and developing univalent foundations in order to bring the languages of mathematics and computer programming closer together. As he saw it, the move to computer formalization is necessary because some branches of mathematics have become too abstract to be reliably checked by people.

"The world of mathematics is becoming very large, the complexity of mathematics is becoming very high, and there is a danger of an accumulation of mistakes," Voevodsky said. Proofs rely on other proofs; if one contains a flaw, all others that rely on it will share the error.

This is something Voevodsky learned through personal experience. In 1999 he discovered an error in a paper he had written seven years earlier. Voevodsky eventually found a way to salvage the result, but in a 2014 article in the IAS newsletter, he wrote that the experience scared him. He began to worry that unless he formalized his work on the computer, he wouldn't have complete confidence that it was correct.

But taking that step required him to rethink the very basics of mathematics. The accepted foundation of mathematics is set theory. Like any foundational system, set theory provides a collection of basic concepts and rules, which can be used to construct the rest of mathematics. Set theory has sufficed as a foundation for more than a century, but it can't readily be translated into a form that computers can use to check proofs. So with his decision to start formalizing mathematics on the computer, Voevodsky set in motion a process of discovery that ultimately led to something far more ambitious: a recasting of the underpinnings of mathematics.

SET THEORY AND A PARADOX

Once the whole numbers are in place, fractions can be defined as pairs of whole numbers, decimals can be defined as sequences of digits, functions in the plane can be defined as sets of ordered pairs and so on. "You end up with complicated structures, which are a set of things, which are a set of things, which are a set of things, all the way down to the metal, to the empty set at the bottom," said Michael Shulman, a mathematician at the University of San Diego.

Set theory as a foundation includes these basic objects—sets—and logical rules for manipulating those sets, from which the theorems in mathematics are derived. An advantage of set theory as a foundational system is that it is very economical—every object mathematicians could possibly want to use is ultimately built from the null set.

On the other hand, it can be tedious to encode complicated mathematical objects as elaborate hierarchies of sets. This limitation becomes problematic when mathematicians want to think about objects that are equivalent or isomorphic in some sense, if not necessarily equal in all respects. For example, the fraction 1/2 and the decimal 0.5 represent the same real number but are encoded very differently in terms of sets.

"You have to build up a specific object and you're stuck with it," Awodey said. "If you want to work with a different but isomorphic object, you'd have to build that up."

But set theory isn't the only way to do mathematics. The proof assistant programs Coq and Agda, for example, are based on a different formal system called type theory.

Type theory has its origins in an attempt to fix a critical flaw in early versions of set theory, which was identified by the philosopher and logician Bertrand Russell in 1901. Russell noted that some sets contain themselves as a member. For example, consider the set of all things that are not spaceships. This set—the set of non-spaceships—is itself not a spaceship, so it is a member of itself.

Russell defined a new set: the set of all sets that do not contain themselves. He asked whether that set contains itself, and he showed that answering that question produces a paradox: If the set does contain itself, then it doesn't contain itself (because the only objects in the set are sets that don't contain themselves). But if it doesn't contain itself, it does contain itself (because the set contains all the sets that don't contain themselves).

Russell created type theory as a way out of this paradox. In place of set theory, Russell's system used more carefully defined objects called types. Russell's type theory begins with a universe of objects, just like set theory, and those objects can be collected in a "type" called a *SET*. Within type theory, the type *SET* is defined so that it is only allowed to collect objects that aren't collections of other things. If a collection does contain other collections, it is no longer allowed to be a *SET*, but is instead something that can be thought of as a *MEGASET*—a new kind of type defined specifically as a collection of objects which themselves are collections of objects.

From here, the whole system arises in an orderly fashion. One can imagine, say, a type called a *SUPERMEGASET* that collects only objects that are *MEGASETS*. Within this rigid framework, it becomes illegal, so to speak, to even ask the paradox-inducing question, "Does the set of all sets that do not contain themselves contain itself?" In type theory, *SETS* only contain objects that are not collections of other objects.

An important distinction between set theory and type theory lies in the way theorems are treated. In set theory, a theorem is not itself a set—it's a statement about sets. By contrast, in some versions of type theory, theorems and *SETS* are on equal footing. They are "types"—a new kind of mathematical object. A theorem is the type whose elements are all the different ways the theorem can be proved. So, for example, there is a single type that collects all the proofs to the Pythagorean theorem.

To illustrate this difference between set theory and type theory, consider two sets: Set A contains two apples and Set B contains two oranges. A mathematician might consider these sets equivalent, or isomorphic, because they have the same number of objects. The way to show formally that these two sets are equivalent is to pair objects from the first set with objects from the second. If they pair evenly, with no objects left over on either side, they're equivalent.

When you do this pairing, you quickly see that there are two ways to show the equivalence: Apple 1 can be paired with Orange 1 and Apple 2 with Orange 2, or Apple 1 can be paired with Orange 2 and Apple 2 with Orange 1. Another way to say this is to state that the two sets are isomorphic to each other in two ways.

In a traditional set theory proof of the theorem Set $A \cong$ Set B (where the symbol $\cong$ means "is isomorphic to"), mathematicians are concerned only with whether one such pairing exists. In type theory, the theorem Set $A \cong$ Set B can be interpreted as a collection, consisting of all the different ways of demonstrating the isomorphism (which in this case is two). There are often good reasons in mathematics to keep track of the various ways in which two objects (like these two sets) are equivalent, and type theory does this automatically by bundling equivalences together into a single type.

This is especially useful in topology, a branch of mathematics that studies the intrinsic properties of spaces, like a circle or the surface of a doughnut. Studying spaces would be impractical if topologists had to think separately about all possible variations of spaces with the same intrinsic properties. (For example, circles can come in any size, yet every circle shares the same basic qualities.) A solution is to reduce the number of distinct spaces by considering some of them to be equivalent. One way topologists do this is with the notion of homotopy, which provides a useful definition of equivalence: Spaces are homotopy equivalent if, roughly speaking, one can be deformed into the other by shrinking or thickening regions, without tearing.

The point and the line are homotopy equivalent, which is another way of saying they're of the same homotopy type. The letter P is of the same homotopy type as the letter O (the tail of the P can be collapsed to a point on the boundary of the letter's upper circle), and both P and O are of the same homotopy type as the other letters of the alphabet that contain one hole—A, D, Q and R.

Topologists use different methods for assessing the qualities of a space and determining its homotopy type. One way is to study the collection of paths between distinct points in the space, and type theory is well-suited to keeping track of these paths. For instance, a topologist might think of two points in a space as equivalent whenever there is a path connecting them. Then the collection of all paths between points x and y can itself be viewed as a single type, which represents all proofs of the theorem $x = y$.

Homotopy types can be constructed from paths between points, but an enterprising mathematician can also keep track of paths between paths, and paths between paths between paths and so on. These paths between paths can be thought of as higher-order relationships between points in a space.

Voevodsky tried on and off for 20 years, starting as an undergraduate at Moscow State University in the mid-1980s, to formalize mathematics in a way that would make these higher-order relationships—paths of paths of paths—easy to work with. Like many others during this period, he tried to accomplish this within the framework of a formal system called category theory. And while he achieved limited success in using category theory to formalize particular regions of mathematics, there remained regions of mathematics that categories couldn't reach.

Voevodsky returned to the problem of studying higher-order relationships with renewed interest in the years after he won the Fields Medal. In late 2005, he had something of an epiphany. As soon as he started thinking about higher-order relationships in terms of objects called infinity-groupoids, he said, "many things started to fall into place."

Infinity-groupoids encode all the paths in a space, including paths of paths, and paths of paths of paths. They crop up in other frontiers of mathematical research as ways of encoding similar higher-order relationships, but they are unwieldy objects from the point of view of set theory. Because of this, they were thought to be useless for Voevodsky's goal of formalizing mathematics.

Yet Voevodsky was able to create an interpretation of type theory in the language of infinity-groupoids, an advance that allows mathematicians to reason efficiently about infinity-groupoids without ever having to think of them in terms of sets. This advance ultimately led to the development of univalent foundations.

Voevodsky was excited by the potential of a formal system built on groupoids, but also daunted by the amount of technical work required to realize the idea. He was also concerned that any progress he made would be too complicated to be reliably verified through peer review, which Voevodsky said he was "losing faith in" at the time.

TOWARD A NEW FOUNDATIONAL SYSTEM

With groupoids, Voevodsky had his object, which left him needing only a formal framework in which to organize them. In 2005 he found it in an unpublished paper called FOLDS, which introduced Voevodsky to a formal system that fit uncannily well with the kind of higher-order mathematics he wanted to practice.

In 1972 the Swedish logician Per Martin-Löf introduced his own version of type theory inspired by ideas from Automath, a formal language for checking proofs on the computer. Martin-Löf type theory (MLTT) was eagerly adopted by computer scientists, who have used it as the basis for proof-assistant programs.

In the mid-1990s, MLTT intersected with pure mathematics when Michael Makkai, a specialist in mathematical logic who retired from McGill University in 2010, realized it might be used to formalize categorical and higher-categorical mathematics. Voevodsky said that when he first read Makkai's work, set forth in FOLDS, the experience was "almost like talking to myself, in a good sense."

Voevodsky followed Makkai's path but used groupoids instead of categories. This allowed him to create deep connections between homotopy theory and type theory.

"This is one of the most magical things, that somehow it happened that these programmers really wanted to formalize [type theory]," Shulman said, "and it turns out they ended up formalizing homotopy theory."

Voevodsky agreed that the connection is magical, though he saw the significance a little differently. To him, the real potential of type theory informed by homotopy theory is as a new foundation for mathematics that's uniquely well-suited both to computerized verification and to studying higher-order relationships.

Voevodsky first perceived this connection when he read Makkai's paper, but it took him another four years to make it mathematically precise. From 2005 to 2009 Voevodsky developed several pieces of machinery that allow mathematicians to work with sets in MLTT "in a consistent and convenient way for the first time," he said. These include a new axiom, known as the univalence axiom, and a complete interpretation of MLTT in the language of simplicial sets, which (in addition to groupoids) are another way of representing homotopy types.

This consistency and convenience reflects something deeper about the program, said Daniel Grayson, an emeritus professor of mathematics at the University of Illinois at Urbana-Champaign. The strength of univalent foundations lies in the fact that it taps into a previously hidden structure in mathematics.

"What's appealing and different about [univalent foundations], especially if you start viewing [it] as replacing set theory," he said, "is that it appears that ideas from topology come into the very foundation of mathematics."

FROM IDEA TO ACTION

Building a new foundation for mathematics is one thing. Getting people to use it is another. By late 2009 Voevodsky had worked out the details of univalent foundations and felt ready to begin sharing his ideas. He understood people were likely to be skeptical. "It's a big thing to say I have something which should probably replace set theory," he said.

Voevodsky first discussed univalent foundations publicly in lectures at Carnegie Mellon in early 2010 and at the Oberwolfach Research Institute for Mathematics in Germany in 2011. At the Carnegie Mellon talks he met Steve Awodey, who had been doing research with his graduate students Michael Warren and Peter Lumsdaine on homotopy type theory. Soon after, Voevodsky decided to bring researchers together for a period of intensive collaboration in order to jump-start the field's development.

Along with Thierry Coquand, a computer scientist at the University of Gothenburg in Sweden, Voevodsky and Awodey organized a special research year to take place at IAS during the 2012–2013 academic year. More than 30 computer scientists, logicians and mathematicians came from around the world to participate. Voevodsky said the ideas they discussed were so strange that at the outset, "there wasn't a single person there who was entirely comfortable with it."

The ideas may have been slightly alien, but they were also exciting. Shulman deferred the start of a new job in order to take part in the project. "I think a lot of us felt we were on the cusp of something big, something really important," he said, "and it was worth making some sacrifices to be involved with the genesis of it."

Following the special research year, activity split in a few different directions. One group of researchers, which includes Shulman and is referred to as the HoTT community (for homotopy type theory), set off to explore the possibilities for new discoveries within the framework they'd developed. Another group, which identifies as UniMath and included Voevodsky, began rewriting mathematics in the language of univalent foundations. Their goal is to create a library of basic mathematical elements—lemmas, proofs, propositions—that mathematicians can use to formalize their own work in univalent foundations.

As the HoTT and UniMath communities have grown, the ideas that underlie them have become more visible among mathematicians, logicians and computer scientists. Henry Towsner, a logician at the University of Pennsylvania, said in 2015 that there seemed to be at least one presentation on homotopy type theory at every conference he attended, and that the more he learned about the approach, the more it made sense. "It was this buzzword," he said. "It took me awhile to understand what they were actually doing and why it was interesting and a good idea, not a gimmicky thing."

A lot of the attention univalent foundations received was due to Voevodsky's standing as one of the greatest mathematicians of his generation. Michael Harris, a mathematician at Columbia University, included a long discussion of univalent foundations in his book, *Mathematics Without Apologies*. He was impressed by the mathematics that surround the univalence model, but he was more skeptical of Voevodsky's larger vision of a world in which all mathematicians formalize their work in univalent foundations and check it on the computer.

"The drive to mechanize proof and proof verification doesn't strongly motivate most mathematicians as far as I can tell," he said. "I can understand why computer scientists and logicians would be excited, but I think mathematicians are looking for something else."

Voevodsky was aware that a new foundation for mathematics is a tough sell, and he admitted that "there is really more hype and noise than the field is ready for." He worked on using the language of univalent foundations to formalize the relationship between MLTT and homotopy theory, which he considered a necessary next step in the development of the field. Voevodsky also planned to formalize his proof of the Milnor conjecture, the achievement for which he earned a Fields Medal. He hoped that doing so might act as "a milestone which can be used to create motivation in the field."

Voevodsky hoped to eventually use univalent foundations to study aspects of mathematics that have been inaccessible within the framework of set theory. But he was approaching the development of univalent foundations cautiously. Set theory has undergirded mathematics for more than a century, and if univalent foundations was to have similar longevity, Voevodsky knew it was important to get things right at the start.

Voevodsky's direct participation in this grand initiative ended abruptly on September 30, 2017, when he died in Princeton, New Jersey, at the age of 51.

LANDMARK ALGORITHM BREAKS 30-YEAR IMPASSE

Erica Klarreich

I n November 2015, a theoretical computer scientist presented an algorithm that was hailed as a breakthrough in mapping the obscure terrain of complexity theory, which explores how hard computational problems are to solve. László Babai, of the University of Chicago, announced that he had come up with a new algorithm for the "graph isomorphism" problem, one of the most tantalizing mysteries in computer science. The new algorithm appeared to be vastly more efficient than the previous best algorithm, which had held the record for more than 30 years. His paper became available the following month on the scientific preprint site arxiv.org, and he also submitted it to the Association for Computing Machinery's 48th Symposium on Theory of Computing.[1]

For decades, the graph isomorphism problem has held a special status within complexity theory. While thousands of other computational problems have meekly succumbed to categorization as either hard or easy, graph isomorphism has defied classification. It seems easier than the hard problems, but harder than the easy problems, occupying a sort of no man's land between these two domains. It is one of the two most famous problems in this strange gray area, said Scott Aaronson, a computer scientist at the University of Texas at Austin. Soon after Babai's announcement, Aaronson said, "it looks as if one of the two may have fallen."

The announcement electrified the theoretical computer science community. If correct, it will be "one of the big results of the decade, if not the last several decades," said Joshua Grochow, a computer scientist at the University of Colorado, Boulder.

Computer scientists use the word "graph" to refer to a network of nodes with edges connecting some of the nodes. The graph isomorphism question simply asks when two graphs are really the same graph in disguise because there's a one-to-one correspondence (an "isomorphism") between their nodes that preserves the ways the nodes are connected. The problem is easy to state, but tricky to solve, since even small graphs can be made to look very different just by moving their nodes around.

Babai took a major step forward in pinning down the problem's difficulty level, by setting forth what he asserts is a "quasi-polynomial-time" algorithm to solve it. As Aaronson described it, the algorithm places the problem within "the greater metropolitan area" of P, the class of problems that can be solved efficiently. While this work was not the final word on how hard the graph isomorphism problem is, researchers saw it as a game changer. "Before his announcement, I don't think anyone, except maybe for him, thought this result was going to happen in the next 10 years, or perhaps even ever," Grochow said.

In late 2015, Babai gave four talks outlining his algorithm. He initially declined to speak to the press, writing in an email: "The integrity of science requires that new results be subjected to thorough review by expert colleagues before the results are publicized in the media."

On January 4, 2016, Babai retracted his claim that the new algorithm runs in quasi-polynomial time. Five days later, he announced that he had fixed the error.

A BLIND TASTE TEST

Given two graphs, one way to check whether they are isomorphic is simply to consider every possible way to match up the nodes in one graph with the nodes in the other. But for graphs with n nodes, the number of different matchings is n factorial ($1 * 2 * 3 * \ldots * n$), which is so much larger than n that this brute-force approach is hopelessly impractical for all but the smallest graphs. For graphs with just 10 nodes, for instance, there are already more than 3 million possible matchings to check. And for graphs with 100 nodes, the number of possible matchings far exceeds the estimated number of atoms in the observable universe.

Computer scientists generally consider an algorithm to be efficient if its running time can be expressed not as a factorial but as a polynomial, such as n^2 or n^3; polynomials grow much more slowly than factorials or exponential functions such as 2^n. Problems that have a polynomial-time algorithm are said to be in the class P, and over the decades since this class was first proposed, thousands of natural problems in all areas of science and engineering have been shown to belong to it.

Computer scientists think of the problems in P as relatively easy, and they think of the thousands of problems in another category, "NP-complete," as hard. No one has ever found an efficient algorithm for an NP-complete problem, and most computer scientists believe no one ever will. The question of whether the NP-complete problems are truly harder than the ones in P is the million-dollar P versus NP problem, widely regarded as one of the most important open questions in mathematics.

The graph isomorphism problem is neither known to be in P nor known to be NP-complete; instead, it seems to hover between the two categories. It is one of only a tiny handful of natural problems that occupy this limbo; the only other such problem that's as well-known as graph isomorphism is the problem of factoring a number into primes. "Lots of people have spent time working on graph isomorphism, because it's a very natural, simple-to-state problem, but it's also so mysterious," Grochow said.

There are good reasons to suspect that graph isomorphism is not NP-complete. For example, it has a strange property that no NP-complete problem has ever been shown to have: It's possible, in theory, for an all-knowing being ("Merlin") to convince an ordinary person ("Arthur") that two graphs are different without giving away any hints about where the differences lie.

This "zero-knowledge" proof is similar to the way Merlin could convince Arthur that Coke and Pepsi have different formulas even if Arthur can't taste the difference between them. All Merlin would have to do is take repeated blind taste tests; if he can always correctly identify Coke and Pepsi, Arthur must accept that the two drinks are different.

Similarly, if Merlin told Arthur that two graphs are different, Arthur could test this assertion by putting the two graphs behind his back, moving their nodes around so that they looked very different from the way they started, and then showing them to Merlin and asking him which was which. If the two graphs are really isomorphic, there's no way Merlin could tell. So if Merlin gets these questions right again and again, Arthur will eventually conclude that the two graphs must be different, even if he can't spot the differences himself.

No one has ever found a blind-taste-test protocol for any NP-complete problem. For that and other reasons, there's a fairly strong consensus among theoretical computer scientists that graph isomorphism is probably not NP-complete.

For the reverse question—whether graph isomorphism is in P—the evidence is more mixed. On the one hand, there are practical algorithms for graph isomorphism that can't solve the problem efficiently for every single graph, but that do well on almost any graph you might throw at them, even randomly chosen ones. Computer scientists have to work hard to come up with graphs that trip these algorithms up.

On the other hand, graph isomorphism is what computer scientists call a "universal" problem: Every possible problem about whether two "combinatorial structures" are isomorphic—for example, the question of whether two different Sudoku puzzles are really the same underlying puzzle—can be recast as a graph isomorphism problem. This means that a fast algorithm for graph isomorphism would solve all these problems at once. "Usually when you have that kind of universality, it implies some kind of hardness," Grochow said.

In 2012, William Gasarch, a computer scientist at the University of Maryland, College Park, informally polled theoretical computer scientists about the graph isomorphism problem and found that 14 people believed it belongs to P, while six believed that it does not. Before Babai's announcement, many people didn't expect the problem to be resolved anytime soon. "I kind of thought maybe it was not in P, or maybe it was but we wouldn't know in my lifetime," Grochow said.

PAINT BY NUMBERS

Babai's proposed algorithm doesn't bring graph isomorphism all the way into P, but it comes close. It is quasi-polynomial, he asserts, which means that for a graph with n nodes, the algorithm's running time is comparable to n raised not to a constant power (as in a polynomial) but to a power that grows very slowly.

The previous best algorithm—which Babai was also involved in creating in 1983 with Eugene Luks, now a professor emeritus at the University of Oregon—ran in "subexponential" time, a running time whose distance from quasi-polynomial time is nearly as big as the gulf between exponential time and polynomial time.[2] Babai, who started working on graph isomorphism in 1977, "has been chipping away at this problem for about 40 years," Aaronson said.

Babai's new algorithm starts by taking a small set of nodes in the first graph and virtually "painting" each one a different color. Then it begins to look for an isomorphism by making an initial guess about which nodes in the second graph might correspond to these nodes, and it paints those nodes the same colors as their corresponding nodes in the first graph. The algorithm eventually cycles through all possible guesses.

Once the initial guess has been made, it constrains what other nodes may do: For example, a node that is connected to the blue node in the first graph must correspond to a node that is connected to the blue node in the second graph. To keep track of these constraints, the algorithm introduces new colors: It might paint nodes yellow if they are linked to a blue node and a red node, or green if they are connected to a red node and two yellow nodes, and so on. The algorithm keeps introducing more colors until there are no connectivity features left to capture.

Once the graphs are colored, the algorithm can rule out all matchings that pair nodes of different colors. If the algorithm is lucky, the painting process will divide the graphs into many chunks of different colors, greatly reducing the number of possible isomorphisms the algorithm has to consider. If, instead, most of the nodes end up the same color, Babai has developed a different way to reduce the number of possible isomorphisms, which works except in one case: when the two graphs contain a structure related to a "Johnson graph." These are graphs that have so much symmetry that the painting process and Babai's further refinements just don't give enough information to guide the algorithm.

In the first of several talks on his new algorithm, on November 10, 2015, Babai called these Johnson graphs "a source of just unspeakable misery" to computer scientists working on painting schemes for the graph isomorphism problem. But Johnson graphs can be handled fairly easily by other methods, so by showing that these graphs are the only obstacle to his painting scheme, Babai was able to tame them.

Babai's approach is "a very natural strategy, very clean in some sense," said Janos Simon, a computer scientist at the University of Chicago. "It's very likely that it's the correct one, but all mathematicians are cautious."

Even though the new algorithm has moved graph isomorphism much closer to P than ever before, Babai speculated in his first talk that the problem may lie just outside its borders, in the suburbs rather than the city center. That would be the most interesting possibility, said Luca Trevisan, a computer scientist at the University of California, Berkeley, since it would make graph isomorphism the first natural problem to have a quasi-polynomial algorithm but no polynomial algorithm. "It would show that the landscape of complexity theory is much richer than we thought," he said. If this is indeed the case, however, don't expect a proof anytime soon: Proving it would amount to solving the P versus NP problem, since it would mean that graph isomorphism separates the problems in P from the NP-complete problems.

Many computer scientists believe, instead, that graph isomorphism is now on a glide path that will eventually send it coasting into P. That is the usual trajectory, Trevisan said, once a quasi-polynomial algorithm has been found. Then again, "somehow this problem has surprised people many times," he said. "Maybe there's one more surprise coming."

A GRAND VISION FOR THE IMPOSSIBLE

Thomas Lin and Erica Klarreich

O ne summer afternoon in 2001, while visiting relatives in India, Subhash Khot drifted into his default mode—quietly contemplating the limits of computation. For hours, no one could tell whether the third-year Princeton University graduate student was working or merely sinking deeper into the living-room couch. That night, he woke up, scribbled something down and returned to bed. Over breakfast the next morning, he told his mother that he had come up with an interesting idea. She didn't know what it was, but her reserved older son seemed unusually happy.

Khot's insight—now called the Unique Games Conjecture—helped him make progress on a problem he was working on at the time, but even Khot and his colleagues did not realize its potential. "It just sounded like an idea that would be nice if it was true," recalled Khot, now a computer science professor at New York University's Courant Institute of Mathematical Sciences.

When Khot returned to Princeton, he mentioned the idea to Sanjeev Arora, his doctoral adviser, who advised him to hold off on publishing it. "I wasn't sure it was going to be a good paper," Arora said. "I thought it was maybe a little premature, that it was just a month since he had the idea."

Khot wrote the paper anyway. "I was just a graduate student," Khot said of the decision. "I wasn't expecting anyone to take me seriously to begin with."[1]

In a sense, Khot's insight completed an idea set in motion by another mentor, Johan Håstad of the Royal Institute of Technology in Stockholm. But even Håstad ignored Khot's conjecture at first. "I thought it might get proven or disproven in a year," he said. "It took us awhile to realize it had all these consequences."

Over the next several years, what seemed a modest observation—that a particular assumption could simplify certain approximation algorithm problems—grew into one of the most influential new ideas in theoretical computer science. In 2014, for his "prescient" assumption and his subsequent leadership in "the effort to understand its complexity and its pivotal role in the study of efficient approximation of optimization problems," Khot was awarded the Rolf Nevanlinna Prize, widely considered one of the top honors in his field.

In announcing the prize, the International Mathematical Union also credited Khot's work for generating "new exciting interactions between computational complexity, analysis and geometry."

Khot, who tends to keep his thoughts close and acknowledgment of his achievements even closer, was as surprised as his colleagues by the success of the Unique Games Conjecture. "I definitely didn't expect that this small proposal would go so far," he said.

Although Arora, like others studying the limits of approximation algorithms, was initially unconvinced by Khot's "pie in the sky" idea, he now credits his former graduate student for sensing that his proposal could clear "a fundamental stumbling block."

"His intuition was right," Arora said. "He's now probably the number one expert in that field."

Assaf Naor, an Israeli mathematician who has worked closely with Khot for almost a decade, nominated his colleague and friend for a $200,000 Microsoft fellowship in 2005, the National Science Foundation's prestigious Alan T. Waterman Award in 2010 and the 2014 Nevanlinna Prize. "I see in his work more than just a collection of really good papers: I see an agenda, an original point of view," Naor said. "There are many talented people who can solve problems—few people can change the way we look at things."

THREE COOKIES

The way Khot looks at things might strike some as pessimistic. Given his research into the theoretical limitations of computers, perhaps it is natural that he tends to see what cannot be done or what might go wrong. When packing for vacations, for example, Khot tries to anticipate any ailments that could strike his 2-year-old son, Neev, by bringing all the medicine he thinks his family could need.

"He has a good sense of what's going to go wrong—he's very analytical," said his wife, Gayatri Ratnaparkhi, in 2014, then an analyst at the Federal Reserve Bank of New York. "And the end result is that not much goes wrong in our lives."

But Khot's analytical approach can also be maddening, Ratnaparkhi said. "He tries to optimize things in every possible way," she said. When walking from point A to point B, for example, Khot always wants to find the short-est path. Ratnaparkhi has to persuade him to take the scenic route. And shopping becomes "a major enterprise," as Khot feels "obliged to go to five stores and take a look at prices," he said. He tries to avoid the outings whenever possible.

Then there were the cookies. One time, inside a local bakery, Khot was surprised to find that three 33-cent cookies were being sold for a dollar. While it didn't prevent him from buying the cookies, he said, "even if it's one cent, it seems off."

Truly, his wife said, "he's in the perfect job for his skill set."

Khot acknowledged that his area of research suits his way of thinking. "In many of the problems that the world faces, in some sense there's an overemphasis on being optimistic," he said. "It's good to know one's own limitations."

Khot's views are also informed by his voracious appetite for other sub-jects, including economics, history and current events. He studies labor sta-tistics, his wife said, and reads seven or eight newspapers a day. "He knows stuff I had no idea he knew," Ratnaparkhi said. "At the museum, looking at Renaissance art, he can tell me what the context is."

While few would mistake Khot's subtle, rimless glasses for the proverbial rose-tinted variety, those who know him best describe him as kind, gentle and giving. "He is a superb adviser and mentor with great graduate students," said Naor, who suggested to NYU's provost in 2007 that the university hire Khot. As Courant Institute colleagues and neighbors in the faculty housing complex, the two have grown closer. "He and his family are my family," Naor said.

Calling Khot a "first-class mathematician," Naor highlighted the impor-tance of the big, abstract questions he studies: "The boundaries between tractable and intractable are inherently interesting to us as humans."

In the three decades preceding Khot's graduate school research, computer scientists had shown that hundreds of important computational challenges belong to a category called "NP-hard" problems, which most computer sci-entists believe cannot be solved exactly by any algorithm that runs in a rea-sonable amount of time. One example is the famous "traveling salesman" problem, which asks for the shortest round-trip route that visits each city in a collection of cities once. By the time Khot arrived at Princeton in 1999, many computer scientists had shifted their focus to exploring efficient algorithms that find good approximate solutions to these difficult problems.

And computer scientists were successful at doing so—for many problems. But in 1992, a team of computer scientists—including Khot's adviser-to-be,

Arora—astonished their colleagues by proving a result called the PCP theorem, which enabled researchers to show that for a wide variety of computational problems, even finding good approximate solutions is NP-hard,
meaning that it's a task that, most computer scientists believe, is impossible
to carry out efficiently.[2]

This revelation dashed researchers' hopes of identifying arbitrarily good
approximation algorithms for every problem, but it opened up a new line
of inquiry: trying to generate "exact hardness" results, statements of the
form, "Here's an approximation algorithm for problem X and a proof that
finding any better approximation is NP-hard." Shortly before Khot started
graduate school, Håstad established exact hardness results for a few approximation problems.[3] It was unclear, however, how to extend his results to
other computation problems.

At Princeton, after breezing through the department's prerequisites in
three months—course work that usually takes good students a year and average students two, Arora said—Khot started playing around with Håstad's
techniques, trying to establish exact hardness results for several problems.
Then came his epiphany while vacationing in India: One of his problems got
much simpler if he made a certain assumption about how difficult a certain
approximation problem is. Back at Princeton, Khot realized that several of
his other problems also became easier if he made the same assumption. He
eventually named this assumption the Unique Games Conjecture.

PROVING THE IMPOSSIBLE

Khot grew up in Ichalkaranji—considered a small town in India with a
population of just under 300,000—where he was well-known for winning
math competitions; his name and picture frequently appeared in the local
Marathi-language newspapers *Sakal* ("Morning") and *Pudhari* ("Leader").
At 16, he achieved the highest score countrywide in the Indian Institute
of Technology Joint Entrance Exam, a test so notoriously difficult that
most eligible students don't bother to take it. At 17, Khot went off to study
computer science at the school in Mumbai without ever having touched a
computer.

Khot was an autodidact from an early age. He loved reading Russian science books that had been translated into Marathi. His favorite one included
chapters describing rare elements like palladium and gallium. But he
seemed destined to follow his parents into the medical profession. His father
and mother—an ear, nose and throat specialist and a general practitioner,
respectively—ran a clinic on the first two floors of the family's residence

that bustled with patients from the local textile industry, many of whom suffered from respiratory ailments.

Then, one day, Khot told his mother that he didn't want to be a doctor. "I was very interested in chemistry and some physics and eventually mathematics," he said. "I kind of realized that math was at the base of all these things; so why not study mathematics?"

This change was for the best, Ratnaparkhi joked. "He would have been a terrible doctor," she said. "He doesn't like to talk to people."

During Khot's high school years, the person who most influenced him was his math teacher and headmaster, V. G. Gogate. "He's like a father or a grandfather," Khot said. "Whenever I go home, he's the first person I would call and the first person I would visit."

After learning in March 2014 that he had won the Nevanlinna Prize, Khot said, the most difficult part of keeping it a secret until the awards ceremony in August was not being able to tell Gogate. When he finds out about the prize, Khot said months before it was announced, "he will be the happiest person, even more so than me or even my mom."

Gogate, now retired, didn't really teach math to Khot, who in his high school years had been learning on his own by reading advanced texts. "There are no good education facilities in our small town," said Khot's mother, Jayashree Khot. "So he had to do it himself."

Gogate invited Khot and other top students to study at his home and encouraged his charges to be self-sufficient and to help others. Khot not only taught himself everything, Gogate said, but he also assisted all of his friends in science, math, Sanskrit, Marathi and English. "He was finding answers to the questions by thinking himself, and he was guiding all his classmates," Gogate said.

But it wasn't easy. Until he began attending IIT, Khot had to figure out the answers to difficult problems by himself. Some problems took him six months to a year to solve. In the end, though, "I think that was very helpful that I learned everything the hard way," he said. Now, Khot believes that his independence and ability to focus are his greatest strengths as a mathematician. "I'm perfectly happy to spend very long periods on a single problem," he said.

About a month into his undergraduate program at IIT, tragedy struck: His father died of a heart attack. "After my father's death, my outlook changed," he said. Test scores and competition standings no longer seemed all-important. He worked hard but worried less about external outcomes.

In Mumbai, Khot completed the required programming homework but gravitated toward the mathematical aspects of computer science, where, he

said, "you don't really need the computer as a machine." When he graduated and went to Princeton, he knew he wanted to focus on theoretical
computer science.

Khot's paper describing the Unique Games Conjecture appeared in
2002.[4] The first hint of the conjecture's power came a year later when Khot
and Oded Regev, now of New York University, showed that if the conjecture is true, then it is possible to establish the exact approximation hardness of a problem about networks called Minimum Vertex Cover.[5] Then,
in 2004, Khot and three collaborators used the conjecture to produce an
unexpected finding.[6] They showed that if the conjecture is true, then the
best known approximation algorithm for another network problem called
Max Cut—an algorithm that many computer scientists had assumed was
just a placeholder until they could find a better one—was truly the best
possible.

Suddenly, everyone was studying the implications of the Unique Games
Conjecture. "You should see the number of mathematicians working on
problems that emanated from this conjecture," said Avi Wigderson of the
Institute for Advanced Study. The years following the Max Cut result wit-
nessed a flood of approximation hardness results—theorems of the form,
"If the Unique Games Conjecture is true, then it's NP-hard to approximate
the solutions of problem X any closer than Y percent."

"This conjecture suddenly became really interesting and important,"
Wigderson said. It even seemed to help prove approximation hardness
results about problems that on the surface seemed to have nothing to do
with the problem at the heart of the Unique Games Conjecture, which
involves assigning colors to the nodes of a network. "What was special
about his problem?" Wigderson asked. "It wasn't clear."

In 2008, Prasad Raghavendra of the University of California, Berkeley,
showed that if the UGC is true, then it's possible to establish the approximation hardness of an entire category of common computational problems
called "constraint satisfaction" problems.[7] These involve trying to find the
solution to a problem that satisfies as many of a set of constraints as possible—for example, the wedding seating chart that places feuding family
members at different tables as much as possible.

"We understand immediately an infinite class of problems by relying
only on the one problem [that Khot] postulated is hard, which is amazing,"
Wigderson said. The conjecture "creates an understanding that one rarely
expects—that's why it's so interesting and beautiful," he said.

"It has changed the way we think about a lot of problems in computer
science," said Ryan O'Donnell, a theoretical computer scientist at Carnegie
Mellon University in Pittsburgh.

TRUE OR FALSE

Khot is most comfortable thinking quietly—whether he's alone in his office, on a Washington Square Park bench surrounded by strollers, buskers and chess hustlers, in a cafe packed with NYU students or at home in India among family and friends.

"When we go to a movie he doesn't like, he's working on his own," Ratnaparkhi said. "That happens quite a lot."

If the Unique Games Conjecture is ever disproved, all the approximation hardness results that emanate from it will collapse. But certain other results will hold firm: For some mysterious reason, the proofs of the approximation hardness results and the attempts to prove the conjecture itself have led mathematicians to state—and then prove—an assortment of theorems about seemingly unrelated areas of mathematics, including the geometry of foams, the relationship between different ways of measuring distance, and even the stability of different election systems.[8] "Out popped these very natural questions," O'Donnell said. These results will hold up regardless of whether the Unique Games Conjecture turns out to be true or false.

It remains to be seen whether computer scientists will be able to prove or disprove Khot's Unique Games Conjecture. A proof would be a boon to computer scientists, but a disproof might be even more exciting, Arora said. Researchers agree that disproving the conjecture would probably require innovative new algorithmic techniques that could unlock a host of different approximation problems. "If someone came up with an efficient algorithm [for the Unique Games problem], we would have a very valuable new insight into how to design algorithms," Arora said.

Khot doesn't expect someone to prove or disprove his conjecture any time soon. "At this point, we can probably hope to just keep constructing pieces of evidence in one direction or another," he said. He is working on proving the conjecture but is also exploring whether he can come up with different avenues toward proving approximation hardness results. "That's the real goal," he said.

Before his son was born, Khot used to think about problems related to the Unique Games Conjecture all the time. But with fatherhood, he said, "you suddenly realize there are much more important things in life than what you thought before."

Heading from his office in 2014—where he had been discussing his work with long pauses and carefully chosen words—to the playground, where he was to pick up his son, Khot could barely hide his anticipation. As the little boy ran to meet him, Khot's brow unfurrowed and a broad smile swept over his face.

"Of all people, he's happiest when he's with Neev," Ratnaparkhi said. "He talks all the time with Neev."

Khot said that wanting to spend more time with his son has made him more efficient at work. Before, he would do his research in between reading the news and browsing the internet. Now, "I have 9 to 5 free to work," Khot said. "I'm much more organized."

Math occasionally creeps in while he's playing with his son, Khot said, but "if the guy is running around, what are you going to do?"

VI | WHAT IS INFINITY?

TO SETTLE INFINITY DISPUTE, A NEW LAW OF LOGIC

Natalie Wolchover

I n the course of exploring their universe, mathematicians have occasionally stumbled across holes: statements that can be neither proved nor refuted with the nine axioms, collectively called "ZFC," that serve as the fundamental laws of mathematics. Most mathematicians simply ignore the holes, which lie in abstract realms with few practical or scientific ramifications. But for the stewards of math's logical underpinnings, their presence raises concerns about the foundations of the entire enterprise.

"How can I stay in any field and continue to prove theorems if the fundamental notions I'm using are problematic?" asked Peter Koellner, a professor of philosophy at Harvard University who specializes in mathematical logic.

Chief among the holes is the continuum hypothesis, a 140-year-old statement about the possible sizes of infinity. As incomprehensible as it may seem, endlessness comes in many measures: For example, there are more points on the number line, collectively called the "continuum," than there are counting numbers. Beyond the continuum lie larger infinities still—an interminable progression of evermore enormous, yet all endless, entities. The continuum hypothesis asserts that there is no infinity between the smallest kind—the set of counting numbers—and what it asserts is the second-smallest—the continuum. It "must be either true or false," the mathematical logician Kurt Gödel wrote in 1947, "and its undecidability from the axioms as known today can only mean that these axioms do not contain a complete description of reality."

The decades-long quest for a more complete axiomatic system, one that could settle the infinity question and plug many of the other holes in mathematics at the same time, arrived at a crossroads in 2013. During a meeting at Harvard organized by Koellner, scholars largely agreed upon two main contenders for additions to ZFC: forcing axioms and the inner-model axiom "V = ultimate L."

"If forcing axioms are right, then the continuum hypothesis is false," Koellner said. "And if the inner-model axiom is right, then the continuum hypothesis is true. You go through a whole list of issues in other fields, and the forcing axioms will answer those questions one way, and ultimate L will answer them a different way."

According to the researchers, choosing between the candidates boils down to a question about the purpose of logical axioms and the nature of mathematics itself. Are axioms supposed to be the grains of truth that yield the most pristine mathematical universe? In that case, V=ultimate L may be most promising.[1] Or is the point to find the most fruitful seeds of mathematical discovery, a criterion that seems to favor forcing axioms?[2] "The two sides have a somewhat divergent view of what the goal is," said Justin Moore, a mathematics professor at Cornell University.

Axiomatic systems like ZFC provide rules governing collections of objects called "sets," which serve as the building blocks of the mathematical universe. Just as ZFC now arbitrates mathematical truth, adding an extra axiom to the rule book would help shape the future of the field—particularly its take on infinity. But unlike most of the ZFC axioms, the new ones "are not self-evident, or at least not self-evident at this stage of our knowledge, so we have a much more difficult task," said Stevo Todorčević, a mathematician at the University of Toronto and the French National Center for Scientific Research in Paris.

Proponents of V=ultimate L say that establishing an absence of infinities between the integers and the continuum promises to bring order to the chaos of infinite sets, of which there are, unfathomably, an infinite variety. But the axiom may have minimal consequences for traditional branches of mathematics.

"Set theory is in the business of understanding infinity," said Hugh Woodin, who is a mathematician at Harvard University; the architect of V=ultimate L; and one of the most prominent living set theorists. The familiar numbers relevant to most mathematics, Woodin argues, "are an insignificant piece of the universe of sets."

Meanwhile, forcing axioms, which deem the continuum hypothesis false by adding a new size of infinity, would also extend the frontiers of mathematics in other directions. They are workhorses that regular mathematicians "can actually go out and use in the field, so to speak," Moore said. "To me, this is ultimately what foundations [of mathematics] should be doing."

Advances in the study of V =ultimate L and newfound uses of forcing axioms, especially one called "Martin's maximum" after the mathematician

Donald Martin, have energized the debate about which axiom to adopt. And there's a third point of view that disagrees with the debate's very premise. According to some theorists, there are myriad mathematical universes, some in which the continuum hypothesis is true and others in which it is false—but all equally worth exploring. Meanwhile, "there are some skeptics," Koellner said, "people who for philosophical reasons think set theory and the higher infinite doesn't even make any sense."

INFINITE PARADOXES

Infinity has ruffled feathers in mathematics almost since the field's beginning. The controversy arises not from the notion of potential infinity—the number line's promise of continuing forever—but from the concept of infinity as an actual, complete, manipulable object.

"What truly infinite objects exist in the real world?" asks Stephen Simpson, a mathematician and logician at Pennsylvania State University. Taking a view originally espoused by Aristotle, Simpson argues that actual infinity doesn't really exist and so it should not so readily be assumed to exist in the mathematical universe. He leads an effort to wean mathematics off actual infinity, by showing that the vast majority of theorems can be proved using only the notion of potential infinity. "But potential infinity is almost forgotten now," Simpson said. "In the ZFC set theory mindset, people tend not to even remember that distinction. They just think infinity means actual infinity and that's all there is to it."

Infinity was boxed and sold to the mathematical community in the late 19th century by the German mathematician Georg Cantor. Cantor invented a branch of mathematics dealing with sets—collections of elements that ranged from empty (the equivalent of the number zero) to infinite. His "set theory" was such a useful language for describing mathematical objects that within decades, it became the field's lingua franca. A nine-item list of rules called Zermelo–Fraenkel set theory with the axiom of choice, or ZFC, was established and widely adopted by the 1920s. Translated into plain English, one of the axioms says two sets are equal if they contain the same elements. Another simply asserts that infinite sets exist.

Assuming actual infinity leads to unsettling consequences. Cantor proved, for instance, that the infinite set of even numbers {2, 4, 6, ... } could be put in a "one-to-one correspondence" with all counting numbers {1, 2, 3, ... }, indicating that there are just as many evens as there are odds-and-evens.

More shocking was his proving in 1873 that the continuum of real numbers (such as 0.00001, 2.568023489, pi and so on) is "uncountable": Real numbers do not correspond in a one-to-one fashion with the counting numbers because for any numbered list of them, it is always possible to come up with a real number that isn't on the list. The infinite sets of real numbers and counting numbers have different sizes, or in Cantor's parlance, different "cardinal numbers." In fact, he found that there are not two but an infinite sequence of ever-larger cardinals, each new infinity consisting of the power set, or set of all subsets, of the infinite set before it.

Some mathematicians despised this mess of infinities. One of Cantor's colleagues called them a "grave disease"; another called him a "corruptor of youth."[3] But by the logic of set theory, it was true.

Cantor wondered about the two smallest cardinals. "It's in some sense the most fundamental question you can ask," Woodin said. "Is there an infinity in between, or is the infinity of the real numbers the first infinity past the infinity of the counting numbers?"

All the obvious candidates for a mid-size infinity fail. Rational numbers (ratios of integers such as 1/2) are countable and thus have the same cardinality as the counting numbers. And there are just as many real numbers in any slice of the continuum (such as between 0 and 1) as there are in the whole set. Cantor guessed that there was no infinity in between countable sets and the continuum. But he couldn't prove this "continuum hypothesis" using the axioms of set theory. Nor could anyone else.

Then, in 1931, Gödel, who had recently finished his doctorate at the University of Vienna, made an astounding discovery. With a pair of proofs, the 25-year-old Gödel showed that a specifiable yet sufficiently complex axiomatic system like ZFC could never be both consistent and complete. Proving that its axioms are consistent (that is, that they don't lead to contradictions) requires an additional axiom not on the list. And to prove that ZFC-plus-that-axiom is consistent, yet another axiom is needed. "Gödel's incompleteness theorems told us we are never going to be able to catch our own tail," Moore said.

The incompleteness of ZFC means that the mathematical universe that its axioms generate will inevitably have holes. "There will be [statements] that cannot be decided by those principles," Woodin said. It soon became clear that the continuum hypothesis, "the most fundamental question you can ask" about infinity, was such a hole. Gödel himself proved that the truth of the continuum hypothesis is consistent with ZFC, and Paul Cohen, an American mathematician, proved the opposite, that the negation of the hypothesis is also consistent with ZFC. Their combined results demonstrated that the continuum hypothesis is actually independent of the axioms. Something beyond ZFC is needed to prove or refute it.

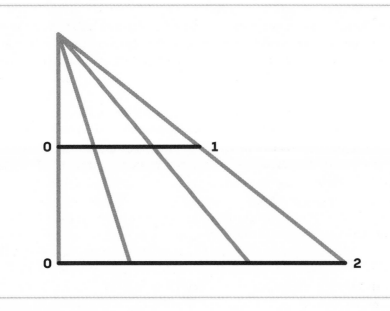

FIGURE 6.1
Because each point on the interval (0, 1) corresponds to the point on (0, 2) that lies on the same gray line, there are just as many real numbers between 0 and 1 as there are between 0 and 2. This "one-to-one correspondence" proves both infinite sets are the same size.

With the hypothesis unresolved, many other properties of cardinal numbers and infinity remain uncertain too. To set theory skeptics like the late Solomon Feferman this didn't matter. "They're simply not relevant to everyday mathematics," Feferman said in 2013.

But to those who spend their days wandering in the universe of sets known as "V," where almost everything is infinite, the questions loom large. "We don't have a clear vision of the universe of sets," Woodin said. "Almost any question you write down about sets is unsolvable. It's not a satisfactory situation."

UNIVERSE OF SETS

Gödel and Cohen, whose combined work led to the current crossroads in set theory, happen to be the founders of the two schools of thought about where to go from here.

Gödel conceived of a small and constructible model universe called "L," populated by starting with the empty set and iterating it to build bigger and bigger sets. In the universe of sets that results, the continuum hypothesis is true: There is no infinite set between that of the integers and the continuum. "Unlike the chaos of the universe of sets, you can really analyze L," Woodin said. This makes the axiom "V=L," or the statement that the universe of sets V is equal to the "inner model" L, appealing. According to Woodin, there's only one problem: "It severely limits the nature of infinity."

L is too small to encompass "large cardinals," infinite sets that ascend in a never-ending hierarchy, with levels named "inaccessible," "measurable," "Woodin," "supercompact," "huge" and so on, altogether composing a cacophonous symphony of infinities. Discovered periodically over the 20th century, these large cardinals cannot be proved to exist with ZFC and instead must be posited with additional "large cardinal axioms." But over the decades, they have been shown to generate rich and interesting mathematics. "As you climb up the large cardinal hierarchy, you get more and more significant consequences," Koellner said.

As many of the mathematicians pointed out, the debate itself reveals a lack of human intuition regarding the concept of infinity.

To keep this symphony of infinities, set theorists have striven for decades to find an inner model that is as pristine and analyzable as L but incorporates large cardinals. However, constructing a universe of sets that included each type of large cardinal required a unique tool kit. For each larger, more inclusive inner model, "you had to do something completely different," Koellner said. "Since the large cardinal hierarchy just goes on and on forever, it looked like we had to go on and on forever too, building as many new inner models as there are transition points in the large cardinal hierarchy. And that kind of makes it look hopeless because, you know, life is short."

Because there was no largest large cardinal, it seemed like there could be no ultimate L, an inner model that encompassed them all. "Then something very surprising happened," Woodin said. In work that was published in 2010, he discovered a breakaway point in the hierarchy.[4]

"Woodin showed that if you can just reach the level of the supercompacts, then there's an overflow and your inner model picks up all the bigger large cardinals as well," Koellner explained. "That was a sort of landscape shift. It provided this new hope that this approach can work. All you have to do is hit one supercompact and then you've got it all."

Although it has not yet been constructed, ultimate L is the name for the hypothetical inner model that includes supercompacts and therefore all large cardinals. The axiom V=ultimate L asserts that this inner model is the universe of sets.

In 2013, Woodin completed the first part of a four-stage proof of the ultimate L conjecture and began vetting it with a small group of colleagues. "It all comes down to this conjecture, and if one can prove it, one proves the existence of ultimate L and verifies it is compatible with all notions of infinity, not only that we have thought of today but that we could ever think of," he said. "If the ultimate L conjecture is true, then there's an absolutely compelling case that V is ultimate L."

EXPANDING THE UNIVERSE

Even if ultimate L exists, can be constructed and is every bit as glorious as Woodin hopes, it isn't everyone's ideal universe. "There's a contrary impulse running through much of set-theoretic history that tells us the universe should be as rich as possible, not as small as possible," said Penelope Maddy, a philosopher of mathematics at the University of California, Irvine and the author of *Defending the Axioms*, published in 2011. "And that's what motivates the forcing axioms."

To expand ZFC, address the continuum hypothesis and better understand infinity, advocates of forcing axioms put stock in a method called forcing, originally conceived of by Cohen. If inner models build a universe of sets from the ground up, forcing expands it outward in all directions.

Todorčević, one of the method's leading specialists, compares forcing to the invention of complex numbers, which are real numbers with an extra dimension. But instead of starting with real numbers, "you are starting with the universe of sets, and then you extend it to form a new, bigger universe," he said. In the extended universe created by forcing, there is a larger class of real numbers than in the original universe defined by ZFC. This means the real numbers of ZFC constitute a smaller infinite set than the full continuum. "In this way, you falsify the continuum hypothesis," Todorčević said.

A forcing axiom called "Martin's maximum," discovered in the 1980s, extends the universe as far as it can go.[5] It is the most powerful rival for V=ultimate L, albeit much less beautiful. "From a philosophical point, it is much harder to justify this axiom," Todorčević said. "It could only be justified in terms of the influence it has on the rest of mathematics."

This is where forcing axioms shine. While V=ultimate L is busy building a castle of unimaginable infinities, forcing axioms fill some problematic potholes in everyday mathematics. Work over recent years by Todorčević,

Moore, Carlos Martinez-Ranero and others showed that they bestow many mathematical structures with nice properties that make them easier to use and understand.[6]

To Moore, these sorts of results give forcing axioms the advantage over inner models. "Ultimately, the decision has to be grounded in: 'What does it do for mathematics?'" he said. "Aside from its own intrinsic interest, what good mathematics does it produce?"

"My response would be, it's certainly true that Martin's maximum is great for understanding structures in classical mathematics," Woodin said. "That's not what set theory is about, to me. It's not clear how Martin's maximum is going to lead to a better understanding of infinity."

At the 2013 Harvard meeting, researchers from both camps presented new work on inner models and forcing axioms and discussed their relative merits. The back-and-forth will likely continue, they said, until one or the other candidate falls by the wayside. Ultimate L could turn out not to exist, for example. Or perhaps Martin's maximum isn't as beneficial as its proponents hope.

As many of the mathematicians pointed out, the debate itself reveals a lack of human intuition regarding the concept of infinity. "Until you further investigate the consequences of the continuum hypothesis, you don't have any real intuition as to whether it's true or false," Moore said.

Mathematics has a reputation for objectivity. But without real-world infinite objects upon which to base abstractions, mathematical truth becomes, to some extent, a matter of opinion—which is Simpson's argument for keeping actual infinity out of mathematics altogether. The choice between $V = $ ultimate L and Martin's maximum is perhaps less of a true-false problem and more like asking which is lovelier, an English garden or a forest?

"It's a personal thing," Moore said.

However, the field of mathematics is known for its unity and cohesion. Just as ZFC came to dominate alternative foundational frameworks in the early 20th century, firmly embedding actual infinity in mathematical thinking and practice, it is likely that only one new axiom to decide the fuller nature of infinity will survive. According to Koellner, "one side is going to have to be wrong."

MATHEMATICIANS BRIDGE FINITE–INFINITE DIVIDE

Natalie Wolchover

W ith a surprising proof in 2016, two young mathematicians found a bridge across the finite–infinite divide, helping at the same time to map this strange boundary.

The boundary does not pass between some huge finite number and the next, infinitely large one. Rather, it separates two kinds of mathematical statements: "finitistic" ones, which can be proved without invoking the concept of infinity, and "infinitistic" ones, which rest on the assumption—not evident in nature—that infinite objects exist.

Mapping and understanding this division is "at the heart of mathematical logic," said Theodore Slaman, a professor of mathematics at the University of California, Berkeley. This endeavor leads directly to questions of mathematical objectivity, the meaning of infinity and the relationship between mathematics and physical reality.

More concretely, the proof settles a question that has eluded top experts for two decades: the classification of a statement known as "Ramsey's theorem for pairs," or RT_2^2. Whereas almost all theorems can be shown to be equivalent to one of a handful of major systems of logic—sets of starting assumptions that may or may not include infinity, and which span the finite–infinite divide—RT_2^2 falls between these lines. "This is an extremely exceptional case," said Ulrich Kohlenbach, a professor of mathematics at the Technical University of Darmstadt in Germany. "That's why it's so interesting."

In the proof, Keita Yokoyama, a mathematician at the Japan Advanced Institute of Science and Technology, and Ludovic Patey, now a computer scientist at CNRS in France, pinned down the logical strength of RT_2^2—but not at a level most people expected.[1] The theorem is ostensibly a statement about infinite objects. And yet, Yokoyama and Patey found that it is "finitistically reducible": It's equivalent in strength to a system of logic that does not invoke infinity. This result means that the infinite apparatus in RT_2^2 can be wielded to prove new facts in finitistic mathematics, forming a surprising bridge between the finite and the infinite. "The result of Patey and Yokoyama is indeed a breakthrough," said Andreas Weiermann of Ghent University in Belgium, whose own work on RT_2^2 unlocked one step of the new proof.

Ramsey's theorem for pairs is thought to be the most complicated statement involving infinity that is known to be finitistically reducible. It invites you to imagine having in hand an infinite set of objects, such as the set of all natural numbers. Each object in the set is paired with all other objects. You then color each pair of objects either red or blue according to some rule. (The rule might be: For any pair of numbers $A < B$, color the pair blue if $B < 2^A$, and red otherwise.) When this is done, RT_2^2 states that there will exist an infinite monochromatic subset: a set consisting of infinitely many numbers, such that all the pairs they make with all other numbers are the same color.

The colorable, divisible infinite sets in RT_2^2 are abstractions that have no analogue in the real world. And yet, Yokoyama and Patey's proof shows that mathematicians are free to use this infinite apparatus to prove statements in finitistic mathematics—including the rules of numbers and arithmetic, which arguably underlie all the math that is required in science—without fear that the resulting theorems rest upon the logically shaky notion of infinity. That's because all the finitistic consequences of RT_2^2 are "true" with or without infinity; they are guaranteed to be provable in some other, purely finitistic way. RT_2^2's infinite structures "may make the proof easier to find," explained Slaman, "but in the end you didn't need them. You could give a kind of native proof—a [finitistic] proof."

When Yokoyama set his sights on RT_2^2 as a postdoctoral researcher six years ago, he expected things to turn out differently. "To be honest, I thought actually it's not finitistically reducible," he said.

This was partly because earlier work proved that Ramsey's theorem for triples, or RT_3^2, is not finitistically reducible: When you color trios of objects in an infinite set either red or blue (according to some rule), the infinite, monochrome subset of triples that RT_3^2 says you'll end up with is too complex an infinity to reduce to finitistic reasoning. That is, compared to the infinity in RT_2^2, the one in RT_3^2 is, so to speak, more hopelessly infinite.

Even as mathematicians, logicians and philosophers continue to parse the subtle implications of Patey and Yokoyama's result, it is a triumph for the "partial realization of Hilbert's program," an approach to infinity championed by the mathematician Stephen Simpson of Vanderbilt University and Pennsylvania State University. The program replaces an earlier, unachievable plan of action by the great mathematician David Hilbert, who in 1921 commanded mathematicians to weave infinity completely into the fold of finitistic mathematics. Hilbert saw finitistic reducibility as the only remedy for the skepticism then surrounding the new mathematics of the infinite. As Simpson described that era, "There were questions about whether mathematics was going into a twilight zone."

THE RISE OF INFINITY

The philosophy of infinity that Aristotle set out in the fourth century B.C. reigned virtually unchallenged until 150 years ago. Aristotle accepted "potential infinity"—the promise of the number line (for example) to continue forever—as a perfectly reasonable concept in mathematics. But he rejected as meaningless the notion of "actual infinity," in the sense of a complete set consisting of infinitely many elements.

Aristotle's distinction suited mathematicians' needs until the 19th century. Before then, "mathematics was essentially computational," said Jeremy Avigad, a philosopher and mathematician at Carnegie Mellon University. Euclid, for instance, deduced the rules for constructing triangles and bisectors—useful for bridge building—and, much later, astronomers used the tools of "analysis" to calculate the motions of the planets. Actual infinity—impossible to compute by its very nature—was of little use. But the 19th century saw a shift away from calculation toward conceptual understanding. Mathematicians started to invent (or discover) abstractions—above all, infinite sets, pioneered in the 1870s by the German mathematician Georg Cantor. "People were trying to look for ways to go further," Avigad said. Cantor's set theory proved to be a powerful new mathematical system. But such abstract methods were controversial. "People were saying, if you're giving arguments that don't tell me how to calculate, that's not math."

And, troublingly, the assumption that infinite sets exist led Cantor directly to some nonintuitive discoveries. He found that infinite sets come in an infinite cascade of sizes—a tower of infinities with no connection to physical reality. What's more, set theory yielded proofs of theorems that were hard to swallow, such as the 1924 Banach–Tarski paradox, which says that if you break a sphere into pieces, each composed of an infinitely dense scattering of points, you can put the pieces together in a different way to create two spheres that are the same size as the original. Hilbert and his contemporaries worried: Was infinitistic mathematics consistent? Was it true?

Amid fears that set theory contained an actual contradiction—a proof of $0=1$, which would invalidate the whole construct—math faced an existential crisis. The question, as Simpson framed it, was, "To what extent is mathematics actually talking about anything real? [Is it] talking about some abstract world that's far from the real world around us? Or does mathematics ultimately have its roots in reality?"

Even though they questioned the value and consistency of infinitistic logic, Hilbert and his contemporaries did not wish to give up such abstractions—power tools of mathematical reasoning that in 1928 would enable the British philosopher and mathematician Frank Ramsey to chop up and color infinite sets at will. "No one shall expel us from the paradise which Cantor has created for us," Hilbert said in a 1925 lecture. He hoped to stay in Cantor's paradise and obtain proof that it stood on stable logical ground. Hilbert tasked mathematicians with proving that set theory and all of infinitistic mathematics is finitistically reducible, and therefore trustworthy. "We must know; we will know!" he said in a 1930 address in Königsberg—words later etched on his tomb.

However, the Austrian-American mathematician Kurt Gödel showed in 1931 that, in fact, we won't. In a shocking result, Gödel proved that no system of logical axioms (or starting assumptions) can ever prove its own consistency; to prove that a system of logic is consistent, you always need another axiom outside of the system. This means there is no ultimate set of axioms—no theory of everything—in mathematics. When looking for a set of axioms that yield all true mathematical statements and never contradict themselves, you always need another axiom. Gödel's theorem meant that Hilbert's program was doomed: The axioms of finitistic mathematics cannot even prove their own consistency, let alone the consistency of set theory and the mathematics of the infinite.

This might have been less worrying if the uncertainty surrounding infinite sets could have been contained. But it soon began leaking into the realm of the finite. Mathematicians started to turn up infinitistic proofs of concrete statements about natural numbers—theorems that could conceivably find applications in physics or computer science. And this top-down reasoning continued. In 1994, Andrew Wiles used infinitistic logic to prove Fermat's Last Theorem, the great number theory problem about which Pierre de Fermat in 1637 cryptically claimed, "I have discovered a truly marvelous proof of this, which this margin is too narrow to contain." Can Wiles' 150-page, infinity-riddled proof be trusted?

With such questions in mind, logicians like Simpson have maintained hope that Hilbert's program can be at least partially realized. Although not all of infinitistic mathematics can be reduced to finitistic reasoning, they argue that the most important parts can be firmed up. Simpson, an adherent of Aristotle's philosophy who has championed this cause since the 1970s (along with Harvey Friedman of The Ohio State University, who first proposed it), estimates that some 85 percent of known mathematical theorems can be reduced to finitistic systems of logic. "The significance of it," he said, "is that our mathematics is thereby connected, via finitistic reducibility, to the real world."

AN EXCEPTIONAL CASE

Almost all of the thousands of theorems studied by Simpson and his followers over the past four decades have turned out (somewhat mysteriously) to be reducible to one of five systems of logic spanning both sides of the finite–infinite divide. For instance, Ramsey's theorem for triples (and all ordered sets with more than three elements) was shown in 1972 to belong at the third level up in the hierarchy, which is infinitistic. "We understood the patterns very clearly," said Henry Towsner, a mathematician at the University of Pennsylvania. "But people looked at Ramsey's theorem for pairs, and it blew all that out of the water."

A breakthrough came in 1995, when the British logician David Seetapun, working with Slaman at Berkeley, proved that RT_2^2 is logically weaker than RT_3^2 and thus below the third level in the hierarchy. The breaking point between RT_2^2 and RT_3^2 comes about because a more complicated coloring procedure is required to construct infinite monochromatic sets of triples than infinite monochromatic sets of pairs.

"Since then, many seminal papers regarding RT_2^2 have been published," said Weiermann—most importantly, a 2012 result by Jiayi Liu (paired with a result by Carl Jockusch from the 1960s) showed that RT_2^2 cannot prove, nor be proved by, the logical system located at the second level in the hierarchy, one rung below RT_3^2. The level-two system is known to be finitistically reducible to "primitive recursive arithmetic," a set of axioms widely considered the strongest finitistic system of logic.[2] The question was whether RT_2^2 would also be reducible to primitive recursive arithmetic, despite not belonging at the second level in the hierarchy, or whether it required stronger, infinitistic axioms. "A final classification of RT_2^2 seemed out of reach," Weiermann said.

But then in January 2016, Patey and Yokoyama, young guns who had been shaking up the field with their combined expertise in computability theory and proof theory, respectively, announced their new result at a conference in Singapore. Using a raft of techniques, they showed that RT_2^2 is indeed equal in logical strength to primitive recursive arithmetic, and therefore finitistically reducible.

"Everybody was asking them, 'What did you do, what did you do?'" said Towsner, who has also worked on the classification of RT_2^2 but said that "like everyone else, I did not get far." He added: "Yokoyama is a very humble guy. He said, 'Well, we didn't do anything new; all we did was, we used the method of indicators, and we used this other technique,' and he proceeded to list off essentially every technique anyone has ever developed for working on this sort of problem."

In one key step, the duo modeled the infinite monochromatic set of pairs in RT_2^2 using a finite set whose elements are "nonstandard" models of the natural numbers. This enabled Patey and Yokoyama to translate the question of the strength of RT_2^2 into the size of the finite set in their model. "We directly calculate the size of the finite set," Yokoyama said, "and if it is large enough, then we can say it's not finitistically reducible, and if it's small enough, we can say it is finitistically reducible." It was small enough.

RT_2^2 has numerous finitistic consequences, statements about natural numbers that are now known to be expressible in primitive recursive arithmetic, and which are thus certain to be logically consistent. Moreover, these statements—which can often be cast in the form "for every number X, there exists another number Y such that..."—are now guaranteed to have primitive recursive algorithms associated with them for computing Y. "This is a more applied reading of the new result," said Kohlenbach. In particular, he said, RT_2^2 could yield new bounds on algorithms for "term rewriting," placing an upper limit on the number of times outputs of computations can be further simplified.

Some mathematicians hope that other infinitistic proofs can be recast in the RT_2^2 language and shown to be logically consistent. A far-fetched example is Wiles' proof of Fermat's Last Theorem, seen as a holy grail by researchers like Simpson. "If someone were to discover a proof of Fermat's theorem which is finitistic except for involving some clever applications of RT_2^2," he said, "then the result of Patey and Yokoyama would tell us how to find a purely finitistic proof of the same theorem."

Simpson considers the colorable, divisible infinite sets in RT_2^2 "convenient fictions" that can reveal new truths about concrete mathematics. But, one might wonder, can a fiction ever be so convenient that it can be thought of as a fact? Does finitistic reducibility lend any "reality" to infinite objects—to actual infinity? There is no consensus among the experts. Avigad is of two minds. Ultimately, he says, there is no need to decide. "There's this ongoing tension between the idealization and the concrete realizations, and we want both," he said. "I'm happy to take mathematics at face value and say, look, infinite sets exist insofar as we know how to reason about them. And they play an important role in our mathematics. But at the same time, I think it's useful to think about, well, how exactly do they play a role? And what is the connection?"

With discoveries like the finitistic reducibility of RT^2_2—the longest bridge yet between the finite and the infinite—mathematicians and philosophers are gradually moving toward answers to these questions. But the journey has lasted thousands of years already and seems unlikely to end anytime soon. If anything, with results like RT^2_2, Slaman said, "the picture has gotten quite complicated."

MATHEMATICIANS MEASURE INFINITIES AND FIND THEY'RE EQUAL

Kevin Hartnett

In a breakthrough that disproved decades of conventional wisdom, two mathematicians showed that two different variants of infinity are actually the same size. The advance touches on one of the most famous and intractable problems in mathematics: whether there exist infinities between the infinite size of the natural numbers and the larger infinite size of the real numbers.

The problem was first identified over a century ago. At the time, mathematicians knew that "the real numbers are bigger than the natural numbers, but not how much bigger. Is it the next biggest size, or is there a size in between?" said Maryanthe Malliaris of the University of Chicago, co-author of the new work along with Saharon Shelah of the Hebrew University of Jerusalem and Rutgers University.

In their new work, Malliaris and Shelah resolve a related 70-year-old question about whether one infinity (call it p) is smaller than another infinity (call it t). They proved the two are in fact equal, much to the surprise of mathematicians.

"It was certainly my opinion, and the general opinion, that p should be less than t," Shelah said.

Malliaris and Shelah published their proof in 2016 in the *Journal of the American Mathematical Society* and were honored in July 2017 with one of the top prizes in the field of set theory.[1] Shelah also went on to win the Rolf Schock prize in 2018. But their work has ramifications far beyond the specific question of how those two infinities are related. It opens an unexpected link between the sizes of infinite sets and a parallel effort to map the complexity of mathematical theories.

MANY INFINITIES

The notion of infinity is mind-bending. But the idea that there can be different sizes of infinity? That's perhaps the most counterintuitive mathematical discovery ever made. It emerges, however, from a matching game even kids could understand.

Suppose you have two groups of objects, or two "sets," as mathematicians would call them: a set of cars and a set of drivers. If there is exactly one driver for each car, with no empty cars and no drivers left behind, then you know that the number of cars equals the number of drivers (even if you don't know what that number is).

In the late 19th century, the German mathematician Georg Cantor captured the spirit of this matching strategy in the formal language of mathematics. He proved that two sets have the same size, or "cardinality," when they can be put into one-to-one correspondence with each other—when there is exactly one driver for every car. Perhaps more surprisingly, he showed that this approach works for infinitely large sets as well.

Consider the natural numbers: 1, 2, 3 and so on. The set of natural numbers is infinite. But what about the set of just the even numbers, or just the prime numbers? Each of these sets would at first seem to be a smaller subset of the natural numbers. And indeed, over any finite stretch of the number line, there are about half as many even numbers as natural numbers, and still fewer primes.

Yet infinite sets behave differently. Cantor showed that there's a one-to-one correspondence between the elements of each of these infinite sets. Because of this, Cantor concluded that all three sets are the same size. Mathematicians call sets of this size "countable," because you can assign one counting number to each element in each set.

1	2	3	4	5	...	(natural numbers)
2	4	6	8	10	...	(evens)
2	3	5	7	11	...	(primes)

After he established that the sizes of infinite sets can be compared by putting them into one-to-one correspondence with each other, Cantor made an even bigger leap: He proved that some infinite sets are even larger than the set of natural numbers.

Consider the real numbers, which are all the points on the number line. The real numbers are sometimes referred to as the "continuum," reflecting their continuous nature: There's no space between one real number and the next. Cantor was able to show that the real numbers can't be put into a one-to-one correspondence with the natural numbers: Even after you create an infinite list pairing natural numbers with real numbers, it's always possible to come up with another real number that's not on your list. Because of this, he concluded that the set of real numbers is larger than the set of natural numbers. Thus, a second kind of infinity was born: the uncountably infinite.

What Cantor couldn't figure out was whether there exists an intermediate size of infinity—something between the size of the countable natural numbers and the uncountable real numbers. He guessed not, a conjecture now known as the continuum hypothesis.

In 1900, the German mathematician David Hilbert made a list of 23 of the most important problems in mathematics. He put the continuum hypothesis at the top. "It seemed like an obviously urgent question to answer," Malliaris said.

In the century since, the question has proved itself to be almost uniquely resistant to mathematicians' best efforts. Do in-between infinities exist? We may never know.

FORCED OUT

Throughout the first half of the 20th century, mathematicians tried to resolve the continuum hypothesis by studying various infinite sets that appeared in many areas of mathematics. They hoped that by comparing these infinities, they might start to understand the possibly non-empty space between the size of the natural numbers and the size of the real numbers.

Many of the comparisons proved to be hard to draw. In the 1960s, the mathematician Paul Cohen explained why. Cohen developed a method called "forcing" that demonstrated that the continuum hypothesis is independent of the axioms of mathematics—that is, it couldn't be proved within the framework of set theory. (Cohen's work complemented work by Kurt Gödel in 1940 that showed that the continuum hypothesis couldn't be disproved within the usual axioms of mathematics.)

Cohen's work won him the Fields Medal in 1966. Mathematicians subsequently used forcing to resolve many of the comparisons between infinities that had been posed over the previous half-century, showing that these too could not be answered within the framework of set theory. (Specifically, Zermelo–Fraenkel set theory plus the axiom of choice.)

Some problems remained, though, including a question from the 1940s about whether p is equal to t. Both p and t are orders of infinity that quantify the minimum size of collections of subsets of the natural numbers in precise (and seemingly unique) ways.

The details of the two sizes don't much matter. What's more important is that mathematicians quickly figured out two things about the sizes of p and t. First, both sets are larger than the natural numbers. Second, p is always less than or equal to t. Therefore, if p is less than t, then p would be an intermediate infinity—something between the size of the natural numbers and the size of the real numbers. The continuum hypothesis would be false.

Mathematicians tended to assume that the relationship between p and t couldn't be proved within the framework of set theory, but they couldn't establish the independence of the problem either. The relationship between p and t remained in this undetermined state for decades. When Malliaris and Shelah found a way to solve it, it was only because they were looking for something else.

AN ORDER OF COMPLEXITY

Around the same time that Paul Cohen was forcing the continuum hypothesis beyond the reach of mathematics, a very different line of work was getting under way in the field of model theory.

For a model theorist, a "theory" is the set of axioms, or rules, that define an area of mathematics. You can think of model theory as a way to classify mathematical theories—an exploration of the source code of mathematics. "I think the reason people are interested in classifying theories is they want to understand what is really causing certain things to happen in very different areas of mathematics," said H. Jerome Keisler, emeritus professor of mathematics at the University of Wisconsin, Madison.

In 1967, Keisler introduced what's now called Keisler's order, which seeks to classify mathematical theories on the basis of their complexity. He proposed a technique for measuring complexity and managed to prove that mathematical theories can be sorted into at least two classes: those that are minimally complex and those that are maximally complex. "It was a small starting point, but my feeling at that point was there would be infinitely many classes," Keisler said.

It isn't always obvious what it means for a theory to be complex. Much work in the field is motivated in part by a desire to understand that question. Keisler describes complexity as the range of things that can happen in a theory—and theories where more things can happen are more complex than theories where fewer things can happen.

A little more than a decade after Keisler introduced his order, Shelah published an influential book, which included an important chapter showing that there are naturally occurring jumps in complexity—dividing lines that distinguish more complex theories from less complex ones. After that, little progress was made on Keisler's order for 30 years.

Then, in her 2009 doctoral thesis and other early papers, Malliaris reopened the work on Keisler's order and provided new evidence for its power as a classification program. In 2011, she and Shelah started working together to better understand the structure of the order. One of their goals was to identify more of the properties that make a theory maximally complex according to Keisler's criterion.

Malliaris and Shelah eyed two properties in particular. They already knew that the first one causes maximal complexity. They wanted to know whether the second one did as well. As their work progressed, they realized that this question was parallel to the question of whether p and t are equal. In 2016, Malliaris and Shelah published a 60-page paper that solved both problems: They proved that the two properties are equally complex (they both cause maximal complexity), and they proved that p equals t.

"Somehow everything lined up," Malliaris said. "It's a constellation of things that got solved."

In July 2017, Malliaris and Shelah were awarded the Hausdorff medal, one of the top prizes in set theory. The honor reflects the surprising, and surprisingly powerful, nature of their proof. Most mathematicians had expected that p was less than t, and that a proof of that inequality would be impossible within the framework of set theory. Malliaris and Shelah proved that the two infinities are equal. Their work also revealed that the relationship between p and t has much more depth to it than mathematicians had realized.

"I think people thought that if by chance the two cardinals were provably equal, the proof would maybe be surprising, but it would be some short, clever argument that doesn't involve building any real machinery," said Justin Moore, a mathematician at Cornell University who has published a brief overview of Malliaris and Shelah's proof.[2]

Instead, Malliaris and Shelah proved that p and t are equal by cutting a path between model theory and set theory that is already opening new frontiers of research in both fields. Their work also finally puts to rest a problem that mathematicians had hoped would help settle the continuum hypothesis. Still, the overwhelming feeling among experts is that this apparently unresolvable proposition is false: While infinity is strange in many ways, it would be almost too strange if there weren't many more sizes of it than the ones we've already found.

VII | IS MATHEMATICS GOOD FOR YOU?

A LIFE INSPIRED BY AN UNEXPECTED GENIUS

John Pavlus

F or the first 27 years of his life, the mathematician Ken Ono was a screw-up, a disappointment and a failure. At least, that's how he saw himself. The youngest son of first-generation Japanese immigrants to the United States, Ono grew up under relentless pressure to achieve academically. His parents set an unusually high bar. Ono's father, an eminent mathematician who accepted an invitation from J. Robert Oppenheimer to join the Institute for Advanced Study, expected his son to follow in his footsteps. Ono's mother, meanwhile, was a quintessential "tiger parent," discouraging any interests unrelated to the steady accumulation of scholarly credentials.

This intellectual crucible produced the desired results—Ono studied mathematics and launched a promising academic career—but at great emotional cost. As a teenager, Ono became so desperate to escape his parents' expectations that he dropped out of high school. He later earned admission to the University of Chicago but had an apathetic attitude toward his studies, preferring to party with his fraternity brothers. He eventually discovered a genuine enthusiasm for mathematics, became a professor and started a family, but fear of failure still weighed so heavily on Ono that he attempted suicide while attending an academic conference. Only after he joined the Institute for Advanced Study himself did Ono begin to make peace with his upbringing.

Through it all, Ono found inspiration in the story of Srinivasa Ramanujan, a mathematical prodigy born into poverty in late-19th-century colonial India. Ramanujan received very little formal schooling, yet he still produced thousands of independent mathematical results, some of which—like the Ramanujan theta function, which has found applications in string theory—are still intensely studied. But despite his genius, Ramanujan's achievements didn't come easily. He struggled to gain acceptance from Western mathematicians and dropped out of university twice before dying of illness at the age of 32.

While Ono, now 50, doesn't compare himself to Ramanujan in terms of ability, he has built his career in part from Ramanujan's insights. In 2014, Ono and his collaborators Michael Griffin and Ole Warnaar published a breakthrough result in algebraic number theory that generalized one of Ramanujan's own results. Ono's work, which is based on a pair of equations called the Rogers–Ramanujan identities, can be used to easily produce algebraic numbers (such as phi, better known as the "golden ratio").

More recently, Ono served as an associate producer and mathematical consultant for *The Man Who Knew Infinity*, a 2015 feature film about Ramanujan's life. And his memoir, *My Search for Ramanujan: How I Learned to Count* (co-authored with Amir D. Aczel), draws connections between Ramanujan's life and Ono's own circuitous path to mathematical and emotional fulfillment.[1] "I wrote this book to show off my weaknesses, to show off my struggles," Ono said. "People who are successful in their careers were not always successful from day one."

Like Ramanujan, who benefited from years of mentorship by the British mathematician G. H. Hardy, Ono credits his own success to serendipitous encounters with teachers who helped his talents flourish. He now spends a great deal of time mentoring his own students at Emory University. Ono has also helped launch the Spirit of Ramanujan Math Talent Initiative, a venture that "strives to find undiscovered mathematicians around the world and match them with advancement opportunities in the field."

Quanta Magazine spoke with Ono in 2016 about finding his way as a mathematician and a mentor, and about Ramanujan's inspiring brand of creativity. An edited and condensed version of the interview follows.

What was so special about Ramanujan's approach to doing mathematics?
First, he was really a poet, not a problem solver. Most professional mathematicians, whether they're in academia or industry, have problems that they're aiming to solve. Somebody wants to prove the Riemann hypothesis, and sets out to do it. That's how we think science should proceed, and in fact almost every scientist should work that way, because in reality science develops through the work of thousands of individuals slowly adding to a body of knowledge. But what you find in Ramanujan's original notebooks is just formula after formula, and it's not apparent where he's going with his ideas. He was someone who could set down the paths of beginnings of important theories without knowing for sure why we would care about them as mathematicians of the future.

He's credited with compiling thousands of identities—that is, equations that are true regardless of what values the variables take. Why is that important?
It is true that the vast majority of the contents of his notebooks are what you would call identities. Identities that relate continued fractions to other functions, expressions for integrals, expressions for hypergeometric functions and expressions for objects that we call q-series.

But that would be a literal interpretation of his notebooks. In my opinion, that would be like taking a cookbook by Julia Child, reading the recipes and saying that it's about assembling chemical compounds into something more complicated. Strictly speaking that would be true, but you would be missing out on what makes delicious recipes so important to us.

Ramanujan's work came through flights of fancy. If he had been asked to explain why he did his work, he would probably say that he recorded formulas that he found beautiful, and they were beautiful because they revealed some unexpected phenomenon. And they're important to us today because these special phenomena that Ramanujan identified, over and over again, have ended up becoming prototypes for big mathematical theories in the 20th and 21st centuries.

Here's an example. In one of his published manuscripts, Ramanujan recorded a lot of elementary-looking results called congruences. In the 1960s, Jean-Pierre Serre, himself a Fields medalist, revisited some of these results, and in them he found glimpses of a theory that he named the theory of Galois representations. This theory of Galois representations is the language that Andrew Wiles used in the 1990s to prove Fermat's last theorem.

There's no "theory of Ramanujan," but he anticipated mathematical structures that would be important to all of these other more contemporary works. He lived 80 years before his time.

How do you approach your own mathematical work—more as an artist, like Ramanujan, or with the aim of solving specific problems, like a scientist?
I'm definitely much more of a scientist. Science proceeds at a much faster rate than when I started in my career in the early 1990s, and I have to stop often to recognize the beauty in it and try not to be so caught up in the more professionalized side of how one does science. The grant getting, the publications, and all of that—I have to admit, I don't like it.

What compelled you to juxtapose your own story with his?
Well, I almost didn't write it. There are a lot of very personal things that I've never told anyone before. It wasn't until I started writing this book that I was mature enough as a parent myself to try to understand the circumstances

that led my parents to raise us the way they did. And as a professor at Emory, I see all these kids under tremendous pressure—rarely pressure that they understand the origin of. So many of these super-talented kids are just going through the motions, and aren't passionate about their studies at all, and that's terrible. I was like that too. I'd given up on ever trying to live up to my parents' expectations, but somehow because I've had Ramanujan as a guardian angel, things have worked out well for me. It makes you a better teacher when you just tell people how hard it was for you.

This book and your story don't fit the typical "great man of science" narrative. I think you'll find that's much more common than people are willing to admit. I didn't discover my passion for mathematics until my early 20s— that's when [my doctoral adviser Basil] Gordon turned me on to mathematics at a time when I didn't think anything was beautiful. I thought it was all about test scores, grades and trying to do as well as possible without putting in effort. Colleges are full of kids who think that way. How do you beat the system, right? I wasn't beating the system. The system was beating me, and Gordon turned me around. When I've told people the story I've discovered that I'm really not alone.

That's what I see in Ramanujan. He was a two-time college dropout whom my father looked up to as a hero—which made no sense to me when I was 16, because I was told I had to be a child prodigy. I was supposed to do my geometry problems during the summer sitting next to my dad while he did his research. I wasn't even really allowed to go out and play, and then to just have my father tell me about Ramanujan out of the blue—it was beyond earth-shattering.

If you'd been interested in something conventionally "artistic," like music, this kind of painful journey toward success would not seem so surprising. Why does it surprise us to hear about a mathematician having the same struggles? For whatever reason, we live in a culture where we think that the abilities of our best scientists and our best mathematicians are somehow just God-given. That either you have this gift or you don't, and it's not related to help, to hard work, to luck. I think that's part of the reason why, when we try to talk about mathematics to the public, so many people just immediately respond by saying, "Well, I was never very good at math. So I'm not really supposed to understand it or identify with it." I might have had some mathematical talent passed through my father genetically, but that was by no means enough. You have to be passionate about a subject.

At the same time, I want it to be known that it's totally OK to fail. In fact, you learn from your mistakes. We learn early on if that you want to be good at playing the violin, you've got to practice. If you want to be good at sports, you practice. But for some crazy reason, our culture assumes that if you're good at math, you were just born with it, and that's it. But you can be so good at math in so many different ways. I failed my [graduate-school] algebra qualifications! That doesn't mean I can't end up being a successful mathematician. But when I tell people I failed at this, nobody believes me.

But Ramanujan seems to be just that: a unique genius who appeared out of nowhere. What does that have to do with a regular person's life?
You think no one can be like Ramanujan? Well, I disagree. I think we can search the world looking for a mathematical talent, just not by the usual metrics. I want teachers and parents to recognize that when you do see unusual talent, instead of demanding that these people have certain test scores, let's find a way to help nurture them. Because I think humanity needs it. I think these are the lessons we learn from Ramanujan.

You're leading the Spirit of Ramanujan Math Talent Initiative. What is this spirit? How do we recognize it?
First of all, it's the idea that talent is often found in the most unforgiving and unpromising of circumstances. It's the responsibility of mentors, teachers and parents first to recognize that talent, which is not always easy to do, and then to offer opportunities that nurture that talent.

There are no age limits, and I don't want this to be a competition where you're recognized for high test scores. I have no trouble finding people who can get an 800 on the math SAT. That's easy. Those people don't need to be identified. They've already self-identified. I'm searching for creativity.

That said, the Spirit of Ramanujan does not require finding the next Ramanujan. We would be super lucky to do that, but if we make opportunities for 30 talented people around the world who are presently working in an intellectual desert, or are subjected to inelastic educational systems where they're not allowed to flourish—or if we can provide an opportunity for someone to work with a scientist who could be their G. H. Hardy—then this initiative will be successful.

Do you wish you had been nurtured differently? Do you resent your parents?
I love my parents. We discussed the draft of the book for months last summer. They were very upset with me at first, because it was difficult for them

to get past the first 30 pages, but now they embrace it. One reviewer actually saw the book as a love letter to my parents and to my mentors, because they taught me skills I needed.

If you had never joined the Institute for Advanced Study, would you still be struggling to reconcile your own path with your parents' expectations? I think I would still be searching for that recognition today if I hadn't gotten there.

Both my parents will tell you that you only get to live once, so you might as well be the very, very best that you can be at whatever you choose. Which I don't necessarily agree with, because if everyone lived that way, there would be nothing but a whole bunch of unhappy people in the world. But that's how they brought us up. They taught me to be competitive. They taught me not to falsely believe I had done well when I hadn't. They taught me standards, and those are important. But it's true that if I hadn't had the opportunity to work at the Institute, I'm not sure I would have been able to write this book. I might still be struggling with these things.

TO LIVE YOUR BEST LIFE, DO MATHEMATICS

Kevin Hartnett

M ath conferences don't usually feature standing ovations, but Francis
Su received one in Atlanta. Su, a mathematician at Harvey Mudd
College in California and the outgoing president of the Mathematical Association of America (MAA), delivered an emotional farewell address at the
2017 Joint Mathematics Meetings of the MAA and the American Mathematical Society in which he challenged the mathematical community to
be more inclusive.

Su opened his talk with the story of Christopher, an inmate serving a long sentence for armed robbery who had begun to teach himself
math from textbooks he had ordered. After seven years in prison, during
which he studied algebra, trigonometry, geometry and calculus, he wrote
to Su asking for advice on how to continue his work. After Su told this
story, he asked the packed ballroom at the Marriott Marquis, his voice
breaking: "When you think of who does mathematics, do you think of
Christopher?"

Su grew up in Texas, the son of Chinese parents, in a town that was predominantly white and Latino. He spoke of trying hard to "act white" as a
kid. He went to college at the University of Texas, Austin, then to graduate
school at Harvard University. In 2015 he became the first person of color
to lead the MAA. In his talk he framed mathematics as a pursuit uniquely
suited to the achievement of human flourishing, a concept the ancient
Greeks called *eudaimonia*, or a life composed of all the highest goods. Su
talked of five basic human desires that are met through the pursuit of mathematics: play, beauty, truth, justice and love.

If mathematics is a medium for human flourishing, it stands to reason
that everyone should have a chance to participate in it. But in his talk Su
identified what he views as structural barriers in the mathematical community that dictate who gets the opportunity to succeed in the field—from the

requirements attached to graduate school admissions to implicit assumptions about who looks the part of a budding mathematician.

When Su finished his talk, the audience rose to its feet and applauded, and many of his fellow mathematicians came up to him afterward to say he had made them cry. A few hours later *Quanta Magazine* sat down with Su in a quiet room on a lower level of the hotel and asked him why he feels so moved by the experiences of people who find themselves pushed away from math. An edited and condensed version of that conversation and a follow-up conversation follows.

The title of your talk was "Mathematics for Human Flourishing." Flourishing is a big idea—what do you have in mind by it?
When I think of human flourishing, I'm thinking of something close to Aristotle's definition, which is activity in accordance with virtue. For instance, each of the basic desires that I mentioned in my talk is a mark of flourishing. If you have a playful mind or a playful spirit, or you're seeking truth, or pursuing beauty, or fighting for justice, or loving another human being—these are activities that line up with certain virtues. Maybe a more modern way of thinking about it is living up to your potential, in some sense, though I wouldn't just limit it to that. If I am loving somebody well, that's living up to a certain potential that I have to be able to love somebody well.

And how does mathematics promote human flourishing?
It builds skills that allow people to do things they might otherwise not have been able to do or experience. If I learn mathematics and I become a better thinker, I develop perseverance, because I know what it's like to wrestle with a hard problem, and I develop hopefulness that I will actually solve these problems. And some people experience a kind of transcendent wonder that they're seeing something true about the universe. That's a source of joy and flourishing.

Math helps us do these things. And when we talk about teaching mathematics, sometimes we forget these larger virtues that we are seeking to cultivate in our students. Teaching mathematics shouldn't be about sending everybody to a Ph.D. program. That's a very narrow view of what it means to do mathematics. It shouldn't mean just teaching people a bunch of facts. That's also a very narrow view of what mathematics is. What we're really doing is training habits of mind, and those habits of mind allow people to flourish no matter what profession they go into.

Several times in your talk you quoted Simone Weil, the French philosopher (and sibling of the famed mathematician André Weil), who wrote, "Every being cries out silently to be read differently." Why did you choose that quote?
I chose it because it says in a very succinct way what the problem is, what causes injustice—we judge, and we don't judge correctly. So "read" means "judged," of course. We read people differently than they actually are.

And how does that apply to the math community?
We do this in lots of different ways. I think part of it is that we have a picture of who actually can succeed in math. Some of that picture has been developed because the only examples we've seen so far are people who come from particular backgrounds. We're not used to, for instance, seeing African-Americans at a math conference, although it's become more and more common.

We're not used to seeing kids from lower socioeconomic backgrounds in college or grad school. So what I was trying to say is: If we're looking for talent, why are we choosing for background? If we really want to have a more diverse set of people in mathematical sciences, we have to take into account the structural barriers that make it hard for people from disadvantaged backgrounds to succeed in math.

We've been hearing more about how these kinds of educational barriers arise in primary and secondary school. Do you argue that they arise in undergraduate and graduate programs as well?
That's right. At every stage we're losing people. So if you look at some of the studies people are doing now about people who take Calculus 1, and how many of them go on to take Calculus 2, you'll find basically that we're losing women and minorities at these critical junctures. This happens for reasons that we can only speculate about. But I'm sure some of it has to do with people in these groups not seeing themselves as belonging in math, possibly because of a negative culture and an unwelcome climate, or because of things that professors or other students are doing to discourage people from continuing.

The obvious problem with this attrition is that when mathematics draws from a smaller pool, we end up with fewer talented mathematicians. But you emphasized in your speech that denying people math is actually denying them an opportunity to flourish.
Math can contribute in a broad way to every person's life whether that person actually becomes a mathematician or not. The goal of broadly getting

people to appreciate math is not at odds with bringing more people into deep mathematics. Connect with people in a deep way and you're going to draw more people into mathematics. Some of them, more of them, are going to go to graduate school, and that will necessarily happen if you address some of these deep desires—for love, truth, beauty, justice, play. If you address some of these deep themes you're going to get more people and a more diverse set of people in deep mathematics.

Some of those desires are easier to relate to math than others. I think people have a somewhat intuitive sense of how a desire for truth or beauty might be realized through math. But you spent a lot of your talk on justice. How does that relate to mathematics?
Justice is a desire that people have, and so it leads to a certain virtue which is to become a just person, somebody who cares about fighting for things that defend basic human dignity. I spent the most time discussing justice in my talk mainly because I feel that our mathematics community can do better; we can become more just. I see a lot of ways in which we can do better and become more virtuous as a community.

Being a mathematician in some ways allows us to see things more for what they are. When people learn not to overgeneralize their arguments, they're going to be very careful not to think that if you're poor you're necessarily uneducated or vice versa. Having a mathematical background certainly helps people to be less governed by their biases.

You've been a successful research mathematician, yet you teach at a small college, Harvey Mudd, that doesn't have a graduate school. That's kind of unusual. Was there a point where you decided you'd prefer to work at a liberal arts college rather than a big research university?
When I was in graduate school at Harvard I realized I loved teaching, and I remember one of my professors from college telling me that the teaching was better at small liberal arts colleges. So when I was on the job market I started looking at those colleges. I was interested in the research track and willing to do that, but I was also very attracted to the liberal arts environment. I chose to go and I love it; I couldn't see myself being anywhere else.

And how do you think working at a liberal arts college shapes the way you look at the mathematics community today?
I think one of the things I didn't address in the talk, but almost did, is the divide in the community between research universities and liberal arts colleges. There is a cultural divide, and the research universities are in some

sense the dominant culture because all of us with Ph.D.s come through research universities. And there's the whole pattern of the dominant culture being completely unaware of what's going on at the liberal arts colleges. So people come up to me and say: "So, you're at Harvey Mudd; are you happy there?" It's almost like assuming I wouldn't be. That happens all the time, so I find it a bit frustrating to feel like I have to say: "No, this is actually my dream job."

What are the consequences of this cultural imbalance?
Well, the downsides are, for instance, that many of the people at research universities would never consider taking students from an undergraduate college. That's the downside; they're missing a lot of talent. So in many ways the issues are analogous to some of the racial issues that are going on.

I think professors at research universities often don't realize that there are a lot of bright kids coming through the liberal arts colleges. What I'm addressing is the very common practice right now in certain graduate schools of only admitting people who've already had a full slate of graduate courses. In other words, they're expecting undergraduates to have taken graduate courses before they even get considered. If you have that kind of structural situation, you are necessarily going to exclude a bunch of people who otherwise might be successful.

One barrier you mentioned in your talk arises when senior professors don't teach introductory classes. Tell me about that.
I'm being a little provocative here as well. I think what that communicates is: "This is not an important enough segment of people for me to put my attention to." I'm certainly not saying everybody who only teaches senior-level courses has this attitude, but I am saying there are a lot of people who think the math major is basically there for the benefit of students who are going to get a Ph.D. That's a problem.

At the Joint Mathematics Meetings there were a number of prizes specifically for women, and a number of women gave invited talks. Has the math community made more progress on gender equality than on racial inclusiveness?
Definitely, racial inclusiveness has not come as far or as fast as gender inclusiveness. Currently about 27 percent of people with Ph.D.s, faculty members, are women, and about 30 percent of the ones who won awards in teaching and service are women. So we're actually doing pretty well on that front. With our writing awards, which are awards for research and exposition—the fraction of women winning those awards is lower.

Can you look at the process by which gender equality has improved and draw any lessons from that about how to improve racial equality in math?
Many of the practices that work to encourage women in math also work for minorities. Part of the issue here is that there just aren't that many minorities who come into college interested in doing STEM majors. So there's something that happened at the secondary and primary school level, and it would help a lot if we could figure out what's going on there.

You used the metaphor of a "secret menu" in Chinese restaurants. What did you mean by that?
If you go to an authentic restaurant in a big city in New York or California, if you are not Chinese they will give you a standard menu that has things in English and Chinese. But if you're Chinese, they'll give you a different menu. Often it's a menu that is written completely in Chinese and has some additional options that aren't on the standard menu. And I think that happens in the math community. If you talk to women and minorities they will often tell you they've had experiences where people discouraged them from going on, either because they don't think a woman should be in math, or for other reasons. So I used the metaphor "secret menu" to mean: Do we have a secret menu? And who gets to look at it?

You told a story about a student who was counseled by a professor to choose a different major on the grounds that the student wasn't good enough to stick with math. Is that common?
I think it's common. Of course we don't have any data, but I've certainly talked to enough people who've had those kinds of experiences to know that it's very frequent and most of those people are women and minorities.

It's been almost a month since you gave your speech, and it's generated a lot of attention on the internet and among mathematicians. What kinds of responses have you received?
Most of the comments have come from people who are grateful to me for mentioning things that haven't necessarily been discussed, but also for identifying some of the deep, underlying things that cause us to do what we do. I think a lot of people, especially women and minorities, have expressed to me how important it was for somebody to say that. We've been having discussions like this in smaller conversations, and a lot of time it's preaching to the choir, and so having somebody say that in a big address at the national meeting I think felt important and helpful to them.

WHY MATH IS THE BEST WAY TO MAKE SENSE OF THE WORLD

Ariel Bleicher

W hen Rebecca Goldin spoke to a recent class of incoming freshmen at George Mason University, she relayed a disheartening statistic: According to a recent study, 36 percent of college students don't significantly improve in critical thinking during their four-year tenure. "These students had trouble distinguishing fact from opinion, and cause from correlation," Goldin explained.

She went on to offer some advice: "Take more math and science than is required. And take it seriously." Why? Because "I can think of no better tool than quantitative thinking to process the information that is thrown at me." Take, for example, the study she had cited. A first glance, it might seem to suggest that a third of college graduates are lazy or ignorant, or that higher education is a waste. But if you look closer, Goldin told her bright-eyed audience, you'll find a different message: "Turns out, this third of students isn't taking any science."

Goldin, a professor of mathematical sciences at George Mason, has made it her life's work to improve quantitative literacy. In addition to her research and teaching duties, she volunteers as a coach at math clubs for elementary- and middle-school students. In 2004, she became the research director of George Mason's Statistical Assessment Service, which aimed "to correct scientific misunderstanding in the media resulting from bad science, politics or a simple lack of information or knowledge." The project has since morphed into STATS (run by the nonprofit Sense About Science USA and the American Statistical Association), with Goldin as its director. Its mission has evolved too: It is now less of a media watchdog and focuses more on education. Goldin and her team run statistics workshops for journalists and have advised reporters at publications including *FiveThirtyEight*, *ProPublica* and *The Wall Street Journal*.

When *Quanta* first reached out to Goldin, she worried that her dual "hats"—those of a mathematician and a public servant—were too "radically different" to reconcile in one interview. In conversation, however, it quickly became apparent that the bridge between these two selves is Goldin's conviction that mathematical reasoning and study is not only widely useful, but also pleasurable. Her enthusiasm for logic—whether she's discussing the manipulation of manifolds in high-dimensional spaces or the meaning of statistical significance—is infectious. "I love, love, love what I do," she said. It's easy to believe her—and to want some of that delight for oneself.

Quanta Magazine spoke with Goldin about finding beauty in abstract thought, how STATS is arming journalists with statistical savvy and why mathematical literacy is empowering. An edited and condensed version of the conversation follows.

Where does your passion for mathematics and quantitative thought come from?

As a young person I never thought I liked math. I absolutely loved number sequences and other curious things that, in retrospect, were very mathematical. At the dinner table, my dad, who is a physicist, would pull out some weird puzzle or riddle that sometimes only took a minute to solve, and other times I'd be like, "Huh, I have no idea how that one works!" But there was an overall framework of joy around solving it.

When did you recognize you could apply that excitement about puzzles to pursuing math professionally?

Actually very late in the game. I was always very strong in math, and I did a lot of math in high school. This gave me the false sense that I knew what math was about: I felt like every next step was a little bit more of the same, just more advanced. It was very clear in my mind that I didn't want to be a mathematician.

But when I went to college at Harvard, I took a course in topology, which is the study of spaces. It wasn't like anything I'd seen before. It wasn't calculus; it wasn't complex calculations. The questions were really complicated and different and interesting in a way I had never expected. And it was just kind of like I fell in love.

You study primarily symplectic and algebraic geometry. How do you describe what you do to people who aren't mathematicians?

One way I might describe it is to say that I study symmetries of mathematical objects. This comes about when you're interested in things like our universe, where the Earth is rotating, and it's also rotating around the sun, and the sun is in a larger system that is rotating. All those rotations are symmetries. There are a lot of other ways symmetries come up, and they can get really, really complicated. So we use neat mathematical objects to think about them, called groups. This is useful because if you're trying to solve equations, and you know you have symmetries, you can essentially find a way mathematically to get rid of those symmetries and make your equations simpler.

What motivates you to study these complex symmetries?
I just think they're really beautiful. A lot of mathematics ultimately is artistic rather than useful. Sometimes you see a picture that's got a lot of symmetry in it, like an M. C. Escher sketch, and it's like, "Wow, that's just so amazing!" But when you study mathematics, you start to "see" things in higher dimensions. You're not necessarily visualizing them in the same way that you could with a sculpture or piece of art. But you start to feel like this whole system of objects that you're looking at, and the symmetries it has, are really just beautiful. There's no other good word.

How did you get involved with STATS?
When I arrived as a professor at George Mason, I knew I wanted to do more than research and mathematics. I love teaching, but I felt like I wanted to do something for the world that was not part of the ivory tower of just solving problems that I thought were really curious and interesting.

When I first joined what became STATS, it was a little bit more "gotcha" work: looking at how the media talks about science and mathematics and pointing out when someone has gotten it wrong. As we've evolved, I've become more and more interested in how journalists think about quantitative issues and how they process them. We found pretty early in our work that there was this huge gap of knowledge and education: Journalists were writing about things that had quantitative content, but they often didn't absorb what they were writing about, and didn't understand it, and didn't have any way to do better because they were often on really tight timelines with limited resources.

So how has your work at STATS changed?
Our mission at STATS has changed to focus on offering journalists two things. One is to be available to answer quantitative questions. They could be as simple as "I don't know how to calculate this percentage," or they could be pretty sophisticated things, like "I've got this data, and I want to apply this model to it, and I just want to make sure that I'm handling the outliers correctly." The other really cool thing that we do is, we go to

individual news agencies and offer workshops on things like confidence intervals, statistical significance, p values and all this highly technical language.

Someone once described to me the advice he gives to journalists. He says, "You should always have a statistician in your back pocket." That's what we hope to be.

What are the most common pitfalls of reporting on statistics?

A favorite one is distinguishing between causation and correlation. People say, "Oh, that's obvious. Of course there's a difference between those two things." But when you get into examples that target our belief system, it's really hard to disassociate them. Part of the problem, I think, is that scientists themselves always want to know more than they can with the tools they have. And they don't always make clear that the questions they're answering aren't necessarily the ones you might think they're answering.

What do you mean?

Like, you might be interested in knowing whether taking hormones is helpful or harmful to women who are postmenopausal. So you start out with a question that's really well-defined: Does it help or hurt? But you can't necessarily answer that question. What you can answer is the question of whether women who take hormones whom you enroll in your study—those specific women—have an increase or decrease in, say, heart disease rates or breast cancer rates or stroke rates compared to a control group or to the general population. But that may not answer your initial question, which is: "Is that going to be the case for me? Or people like me? Or the population as a whole?"

What do you hope STATS will achieve?

Partly our goal is to help change the culture of journalism so that people recognize the importance of using quantitative arguments and thinking about quantitative issues before they come to conclusions. That way, they're coming to conclusions that are supported by science rather than using a study to further their own agenda—which is something scientists do too; they may push a certain interpretation of something. We want to arm journalists with a certain amount of rigor in their thinking so they can challenge a scientist who might say, "Well, you just don't understand my sophisticated statistic." There's a lot of value in giving reporters the tools to develop their sense of quantitative skepticism so that they're not just bullied.

You argue that statistical literacy gives citizens a kind of power. What do you mean?

What I mean is that if we don't have the ability to process quantitative information, we can often make decisions that are more based on our beliefs and our fears than based on reality. On an individual level, if we have the ability to think quantitatively, we can make better decisions about our own health, about our own choices with regard to risk, about our own lifestyles. It's very empowering to not be scared or bullied into doing things one way or another.

On a collective level, the impact of being educated in general is huge. Think about what democracy would be if most of us couldn't read. We aspire to a literate society because it allows for public engagement, and I think this is also true for quantitative literacy. The more we can get people to understand how to view the world in a quantitative way, the more successful we can be at getting past biases and beliefs and prejudices.

You've also said that getting people to understand statistics requires more than reciting numbers. Why do you think storytelling is important for conveying statistical concepts?

As human beings, we live in stories. It doesn't matter how quantitative you are, we're all influenced by stories. They become like statistics in our mind. So if you report the statistics without the story, you don't get nearly the level of interest or emotion or willingness to engage with the ideas.

How has the media's use of data changed in the 13 years you've been with STATS?

With the internet, we see a tremendous growth in data produced by search engines. Journalists are becoming much more adept at collecting these kinds of data and using them in media articles. I think that the current president is also causing a lot of reflection on what we mean by facts, and in that sense journalists maybe think of it as more important in general to get the facts right.

That's interesting. So you think the public's awareness of "fake" news and "alternative" facts is motivating journalists to be more rigorous about fact checking?

I do think it's very motivating. Of course sometimes information gets spun. But ultimately a very small percentage of journalists do that. I think 95 percent of both journalists and scientists are really working hard to get it right.

I'm surprised you're not more jaded about the media.
Ha! This is maybe more a life view. I think there are people who are pessimistic about humankind and people who are optimistic.

You also volunteer with math clubs for kids. What ideas about math and math culture do you try to get across?
I try to bring in problems that are really different and fun and curious and weird. For example, I've done an activity with kids where I've brought in a bunch of ribbons, and I had them learn a little bit about a field called knot theory. There are two things I'm trying to get across to them. One is that math in school is not the whole story—there's this whole other world that is logical but also beautiful and creative. The second message is a certain emotional framework that I have to offer: that math is a joyous experience.

ACKNOWLEDGMENTS

A publication is only as good as the people who produce it—people, plural, because little of value in journalism or publishing is achieved as a solitary act. First, I give my heartfelt thanks to the many tremendously talented writers, editors and artists who crafted the words and pictures in this book, breathing life into these wonderfully illuminating math stories. I especially want to acknowledge senior writers Kevin Hartnett and Natalie Wolchover as well as frequent contributor Erica Klarreich for their many contributions to this collection.

In addition to the bylined authors, I wish to thank my esteemed magazine coeditors, Michael Moyer and John Rennie, for graciously and intelligently reviewing story ideas, assigning articles, guiding writers and safeguarding *Quanta*'s standards; art director Olena Shmahalo for her sublime vision that permeates the magazine's visual identity; graphics editor Lucy Reading-Ikkanda for transforming impossibly abstract concepts into elegant, accessible visualizations; contributing artist Sherry Choi for retooling the graphics for the book in a beautiful and consistent style; artist Filip Hodas for creating the imaginative covers; unsung heroes like Roberta Klarreich and all of our contributing copy editors for cleaning and polishing our prose, and Matt Mahoney and all of our contributing fact-checkers for acting as a last line of defense and allowing me to sleep better at night; Molly Frances for her meticulous formatting of the reference notes; and producers Jeanette Kazmierczak and Michelle Yun for doing the little things without which everything would come to a grinding halt.

Quanta Magazine, and these books by extension, would not exist without the generous support of the Simons Foundation. I would like to express my deepest gratitude to foundation leaders Jim and Marilyn Simons and Marion Greenup for believing in this project and nurturing it with kindness, wisdom and intellectual rigor every step of the way. I thank Stacey

Greenebaum for her creative public outreach efforts, Jennifer Maimone-Medwick and Yolaine Seaton for their conscientious reviewing of contract language, the entire *Quanta* team for being the best in the business, our distinguished advisory board members for their invaluable counsel and our wonderful foundation colleagues—too many to name individually—for making our work lives both easier and more entertaining.

Special thanks to foreword writer James Gleick, who graced this book with his deep intellect and artful prose in sharing his insights from decades at the forefront of science writing.

I'd like to thank the amazing team at the MIT Press, starting with acquisitions editor Jermey Matthews and book designer Yasuyo Iguchi, who have been a delight to work with, and with a special shout-out to director Amy Brand, who first reached out to me about publishing a *Quanta* book and who provided the leadership, enthusiasm and resources that allowed this project to thrive.

Producing a book is like assembling a big Rube Goldberg machine with countless moving parts and endless opportunities for mistakes and failure. I'm lucky to have found book agent Jeff Shreve of the Science Factory, whose wise feedback and guidance played no small part in averting any number of missteps and making this book a reality.

I'm grateful to the scientists and mathematicians who answered calls from our reporters, editors and fact checkers and patiently and sure-footedly guided us through treacherous territory filled with technical land mines.

I owe everything to my parents, David and Lydia, who gifted me with a lifelong appreciation for science and math; my brother, Ben, who is an inspiration as a high school math teacher; and my wife, Genie, and sons, Julian and Tobias, who give life infinite meaning.

—Thomas Lin

CONTRIBUTORS

Ariel Bleicher is a science writer based in New York City. Her work has appeared in *Quanta Magazine, Scientific American, Discover* and other publications. She was formerly an editor at *Nautilus* and *IEEE Spectrum*.

Robbert Dijkgraaf is director and Leon Levy Professor at the Institute for Advanced Study in Princeton, New Jersey. He is an author, with Abraham Flexner, of *The Usefulness of Useless Knowledge*.

Kevin Hartnett is a senior writer at *Quanta Magazine* covering mathematics and computer science. His work has been collected in the *Best Writing on Mathematics* series in 2013, 2016 and 2017. From 2013 to 2016 he wrote "Brainiac", a weekly column for the *Boston Globe*'s Ideas section.

Erica Klarreich has been writing about mathematics and science for more than 15 years. She has a doctorate in mathematics from Stony Brook University and is a graduate of the Science Communication Program at the University of California, Santa Cruz. Her work has been reprinted in the 2010, 2011 and 2016 volumes of *The Best Writing on Mathematics*.

Thomas Lin is the founding editor-in-chief of *Quanta Magazine*. He previously managed the online science and national news sections at the *New York Times*, where he won a White House News Photographers Association's "Eyes of History" award and wrote about science, tennis and technology. He has also written for the *New Yorker, Tennis Magazine* and other publications.

John Pavlus is a writer and filmmaker whose work has appeared in *Quanta Magazine, Scientific American, Bloomberg Businessweek* and *The Best American Science and Nature Writing* series. He lives in Portland, Oregon.

Siobhan Roberts is a science writer based in Toronto. Her most recent book is *Genius at Play: The Curious Mind of John Horton Conway*.

Natalie Wolchover is a senior writer at *Quanta Magazine* covering the physical sciences. Her work has been featured in *The Best Writing on Mathematics* and recognized with the 2016 Evert Clark/Seth Payne Award and the 2017 American Institute of Physics Science Writing Award. She studied graduate-level physics at the University of California, Berkeley.

NOTES

Unheralded Mathematician Bridges the Prime Gap

1. H. A. Helfgott, "Major Arcs for Goldbach's Theorem" (May 13, 2013), https://arxiv.org/abs/1305.2897v1.pdf.

2. D. A. Goldston, J. Pintz and C. Y. Yildirim, "Primes in Tuples I" (August 10, 2005), https://arxiv.org/abs/math/0508185.

3. Gerald Alexanderson, David F. Hayes and Tatiana Shubin, eds., *Expeditions in Mathematics* (Washington, D.C.: Mathematical Association of America, 2011), https://books.google.com/books?id=DfRWtmWs3hcC&pg=PA101&lpg=PA101&dq=%22level+of+distribution%22&source=bl&ots=jwmsaaSR17&sig=9kfkf6phL66tuisui9BmeaaAbaw&hl=en&sa=X&ei=NhmZUfTBC6fj4APrjIG4AQ.

4. E. Bombieri, J. B. Friedlander and H. Iwaniec, "Primes in Arithmetic Progressions to Large Moduli," *Acta Mathematica* 156, no. 1 (July 1986): 203–251, https://link.springer.com/article/10.1007/BF02399204.

Together and Alone, Closing the Prime Gap

1. James Maynard, "Small Gaps between Primes" (November 20, 2013), https://arxiv.org/abs/1311.4600.

2. "Sieving Admissible Tuples," https://math.mit.edu/~primegaps/sieve.html?ktuple=632.

3. Daniel Goldston, János Pintz and Cem Yildirim, "Primes in Tuples I," *Annals of Mathematics* 170, no. 2 (2009): 819–86.

Mathematicians Discover Prime Conspiracy

1. Robert J. Lemke Oliver and Kannan Soundararajan, "Unexpected Biases in the Distribution of Consecutive Primes" (May 30, 2016), https://arxiv.org/abs/1603.03720.

2. Harald Cramér, "On the Order of Magnitude of the Difference between Consecutive Prime Numbers," *Acta Arithmetica 2* (1937): 23–46, http://matwbn.icm.edu.pl /ksiazki/aa/aa2/aa212.pdf.

3. G. H. Hardy and J. E. Littlewood, "Some Problems of 'Partitio Numerorum'; III: On the Expression of a Number as a Sum of Primes," *Acta Mathematica* 44, no. 1 (1923), https://link.springer.com/article/10.1007%2FBF02403921.

Mathematicians Chase Moonshine's Shadow

1. J. H. Conway and S. P. Norton, "Monstrous Moonshine," *Bulletin of the London Mathematical Society* 11, no. 3 (October 1, 1979): 308–339, https://academic.oup .com/blms/article/11/3/308/339059.

2. Robert L. Griess Jr., "The Friendly Giant," *Inventiones Mathematicae* 69, no. 1 (February 1982): 1–102, https://link.springer.com/article/10.1007%2FBF01389186.

3. Richard E. Borcherds, "Monstrous Moonshine and Monstrous Lie," *Inventiones Mathematicae* 109, no. 1 (December 1992): 405–444, https://link.springer.com/article /10.1007%2FBF01232032.

4. John F. R. Duncan, Michael J. Griffin and Ken Ono, "Proof of the Umbral Moonshine Conjecture," *Research in the Mathematical Sciences* 2, no. 26 (December 15, 2015), https://arxiv.org/abs/1503.01472; Miranda C. N. Cheng, John F. R. Duncan and Jeffrey A. Harvey, "Umbral Moonshine," *Communications in Number Theory and Physics* 8, no. 2 (2014): 101–242, https://arxiv.org/abs/1204.2779.

5. Andrew Wiles, "Modular Elliptic Curves and Fermat's Last Theorem," *Annals of Mathematics* 141, no. 3 (May 1995): 443–551, http://www.jstor.org/stable/2118559 ?origin=crossref&seq=1#page_scan_tab_contents.

6. Tohru Eguchi, Hirosi Ooguri and Yuji Tachikawa, "Notes on the K3 Surface and the Mathieu Group M_24," *Experimental Mathematics* 20, no. 1 (2011): 91–96, https:// arxiv.org/abs/1004.0956.

7. S. P. Zwegers, "Mock Theta Functions" (2002), https://dspace.library.uu.nl/handle /1874/878.

8. Duncan, Griffin and Ono, "Proof of the Umbral Moonshine Conjecture," https:// arxiv.org/abs/1503.01472.

9. Edward Witten, "Three-Dimensional Gravity Revisited" (June 22, 2007), https:// arxiv.org/pdf/0706.3359.pdf.

In Mysterious Pattern, Math and Nature Converge

1. H. L. Montgomery, "The Pair Correlation of Zeros of the Zeta Function," http://www-personal.umich.edu/~hlm/paircor1.pdf.

2. Milan Krbálek and Petr Seba, "The Statistical Properties of the City Transport in Cuernavaca (Mexico) and Random Matrix Ensembles," *Journal of Physics A* 33, no. 26 (July 7, 2000), http://iopscience.iop.org/0305-4470/33/26/102.

3. László Erdős et al., "Spectral Statistics of Erdős–Rényi Graphs I: Local Semicircle Law" (November 9, 2011), https://arxiv.org/pdf/1103.1919v4.pdf.

4. N. Benjamin Murphy and Kenneth M. Golden, "Random Matrices, Spectral Measures and Composite Media" (September 20, 2012), http://jointmathematicsmeetings.org/amsmtgs/2141_abstracts/1086-35-1278.pdf.

At the Far Ends of a New Universal Law

1. Robert M. May, "Will a Large Complex System Be Stable? *Nature* 238 (August 18, 1972): 413–414, http://www.nature.com/nature/journal/v238/n5364/abs/238413a0.html.

2. Terence Tao and Van Vu, "Random Matrices: The Universality Phenomenon for Wigner Ensembles" (February 1, 2012), https://arxiv.org/abs/1202.0068.

3. Amir Aazami and Richard Easther, "Cosmology from Random Multifield Potentials," *Journal of Cosmology and Astroparticle Physics* 2006 (March 2006), https://arxiv.org/pdf/hep-th/0512050v2.pdf.

4. David S. Dean and Satya N. Majumdar, "Large Deviations of Extreme Eigenvalues of Random Matrices," *Physical Review Letters* 97, no. 16 (October 20, 2006), https://journals.aps.org/prl/abstract/10.1103/PhysRevLett.97.160201.

5. David J. Gross and Edward Witten, "Possible Third-Order Phase Transition in the Large-N Lattice Gauge Theory," *Physical Review D* 21, no. 2 (January 15, 1980): 446, https://journals.aps.org/prd/abstract/10.1103/PhysRevD.21.446.

6. Satya N. Majumdar and Grégory Schehr, "Top Eigenvalue of a Random Matrix: Large Deviations and Third Order Phase Transition," *Journal of Statistical Mechanics: Theory and Experiment* 2014 (January 2014), http://iopscience.iop.org/1742-5468/2014/1/P01012?rel=ref&relno=1.

7. Pasquale Calabrese and Pierre Le Doussal, "Exact Solution for the Kardar-Parisi-Zhang Equation with Flat Initial Conditions," *Physical Review Letters* 106, no. 25 (June 24, 2011), https://arxiv.org/pdf/1104.1993.pdf.

8. Kazumasa A. Takeuchi and Masaki Sano, "Universal Fluctuations of Growing Interfaces: Evidence in Turbulent Liquid Crystals," *Physical Review Letters* 104, no. 23 (June 11, 2010), https://journals.aps.org/prl/abstract/10.1103/PhysRevLett.104 .230601.

A Bird's-Eye View of Nature's Hidden Order

1. Yang Jiao et al., "Avian Photoreceptor Patterns Represent a Disordered Hyperuniform Solution to a Multiscale Packing Problem," *Physical Review E* 89, no. 2 (February 24, 2014), https://journals.aps.org/pre/abstract/10.1103/PhysRevE.89.022721; Andrea Gabrielli, Michael Joyce and Francesco Sylos Labini, "Glass-Like Universe: Real-Space Correlation Properties of Standard Cosmological Models," *Physical Review D* 65, no. 8 (April 11, 2002), https://journals.aps.org/prd/abstract/10.1103/PhysRevD .65.083523.

2. Salvatore Torquato and Frank H. Stillinger, "Local Density Fluctuations, Hyperuniformity and Order Metrics," *Physical Review E* 68, no. 4 (October 29, 2003); [Erratum, *Phys. Rev. E* 68, no. 6 (December 15, 2003)], https://journals.aps.org/pre /abstract/10.1103/PhysRevE.68.041113.

3. Joost H. Weijs et al., "Emergent Hyperuniformity in Periodically Driven Emulsions," *Physical Review Letters* 115, no. 10 (September 4, 2015), https://journals.aps .org/prl/abstract/10.1103/PhysRevLett.115.108301.

4. Ludovic Berthier et al., "Suppressed Compressibility at Large Scale in Jammed Packings of Size-Disperse Spheres," *Physical Review Letters* 106, no. 12 (March 21, 2011), https://journals.aps.org/prl/abstract/10.1103/PhysRevLett.106.120601.

5. Olivier Leseur, Romain Pierrat and Rémi Carminati, "High-Density Hyperuniform Materials Can Be Transparent" (May 13, 2016), https://arxiv.org/pdf/1510.05807v3 .pdf.

6. Weining Man et al., "Isotropic Band Gaps and Freeform Waveguides Observed in Hyperuniform Disordered Photonic Solids," *Proceedings of the National Academy of Sciences* 110, no. 40 (October 2013): 15886–15891, http://physics.princeton.edu /~steinh/PNAS-2013-Man-15886-91.pdf.

A Unified Theory of Randomness

1. Jason Miller and Scott Sheffield, "Liouville Quantum Gravity and the Brownian Map I: The QLE(8/3,0) Metric" (February 27, 2016), https://arxiv.org/abs/1507.00719; Jason Miller and Scott Sheffield, "Liouville Quantum Gravity and the Brownian Map II: Geodesics and Continuity of the Embedding" (May 11, 2016), https://arxiv.org/ abs/1605.03563.

2. A. M. Polyakov, "Quantum Geometry of Bosonic Strings," *Physics Letters B* 103, no. 3 (July 23, 1981): 207–210, http://www.sciencedirect.com/science/article/pii /0370269381907437.

3. Miller and Sheffield, "Liouville Quantum Gravity and the Brownian Map I," https://arxiv.org/abs/1507.00719; Miller and Sheffield, "Liouville Quantum Gravity and the Brownian Map II," https://arxiv.org/abs/1605.03563.

Strange Numbers Found in Particle Collisions

1. D. J. Broadhurst and D. Kreimer, "Knots and Numbers in φ^4 Theory to 7 Loops and Beyond," *International Journal of Modern Physics C* 6, no. 4 (August 1995), https:// arxiv.org/abs/hep-ph/9504352.

2. Francis Brown and Oliver Schnetz, "A K3 in φ^4," *Duke Mathematical Journal* 161, no. 10 (2013): 1817–1862, https://projecteuclid.org/euclid.dmj/1340801625.

Quantum Questions Inspire New Math

1. Philip Candelas et al., "A Pair of Calabi–Yau Manifolds as an Exactly Soluble Superconformal Theory," *Nuclear Physics B* 359, no. 1 (July 29, 1991): 21–74, https:// www.sciencedirect.com/science/article/pii/0550321391902926.

A Path Less Taken to the Peak of the Math World

1. Karim Adiprasito, June Huh and Eric Katz, "Hodge Theory for Combinatorial Geometries" (November 9, 2015), https://arxiv.org/abs/1511.02888.

2. William P. Thurston, "On Proof and Progress in Mathematics," *Bulletin of the American Mathematical Society* 30, no. 2 (April 1, 1994): 161–177, https://arxiv.org /abs/math/9404236.

3. Adiprasito, Huh and Katz, "Hodge Theory for Combinatorial Geometries," https:// arxiv.org/abs/1511.02888.

A Long-Sought Proof, Found and Almost Lost

1. T. Royen, "A Simple Proof of the Gaussian Correlation Conjecture Extended to Multivariate Gamma Distributions" (August 13, 2014), https://arxiv.org/pdf/1408 .1028.pdf.

2. Rafał Latała and Dariusz Matlak, "Royen's Proof of the Gaussian Correlation Inequality," in *Geometric Aspects of Functional Analysis. Lecture Notes in Mathematics, Vol. 2169,* ed. B. Klartag and E. Springer (Cham, Switzerland: Springer, 2015), https:// arxiv.org/pdf/1512.08776.pdf.

3. S. Das Gupta et al., "Inequalities on the Probability Content of Convex Regions for Elliptically Contoured Distributions," in *Proceedings of the Sixth Berkeley Symposium on Mathematical Statistics and Probability,* ed. Lucien Le Cam, Jerzy Neyman and Elizabeth L. Scott (University of California Press, 1972), https://books.google.com/books?hl=en&lr=&id=q_QPPufvfuQC&oi=fnd&pg=PA241&dq=gaussian+correlation+inequality+1972+das+gupta&ots=edbLltuP58&sig=13YNjoo4zlmfaslJL78YKIK9N_s.

4. Loren D. Pitt, "A Gaussian Correlation Inequality for Symmetric Convex Sets," *Annals of Probability* 5, no. 3 (1977): 470–474, https://projecteuclid.org/euclid.aop/1176995808.

5. Olive Jean Dunn, "Estimation of the Medians for Dependent Variables," *Annals of Mathematical Statistics* 30, no. 1 (March 1959): 192–197, https://www.jstor.org/stable/2237135?seq=1#page_scan_tab_contents.

"Outsiders" Crack 50-Year-Old Math Problem

1. Peter G. Casazza, "Consequences of the Marcus/Spielman/Stivastava Solution to the Kadison–Singer Problem" (January 5, 2015), https://arxiv.org/pdf/1407.4768.pdf.

2. Adam Marcus, Daniel A Spielman and Nikhil Srivastava, "Interlacing Families II: Mixed Characteristic Polynomials and the Kadison–Singer Problem" (April 14, 2014), https://arxiv.org/abs/1306.3969.

3. Richard V. Kadison and I. M. Singer, "Extensions of Pure States," *American Journal of Mathematics* 81, no. 2 (April 1959): 383–400, http://www.jstor.org/stable/2372748?seq=1#page_scan_tab_contents.

4. Joel Anderson, "Extensions, Restrictions and Representations of States on C*-Algebras," *Transactions of the American Mathematical Society* 249, no. 2 (February 1979): 303–329, http://www.ams.org/journals/tran/1979-249-02/S0002-9947-1979-0525675-1/S0002-9947-1979-0525675-1.pdf.

5. Peter G. Casazza and Janet Crandell Tremain, "The Kadison–Singer Problem in Mathematics and Engineering," *Proceedings of the National Academy of Sciences* 103, no. 7 (February 2006): 2032–2039, https://arxiv.org/pdf/math/0510024v2.pdf.

6. Nik Weaver, "The Kadison–Singer Problem in Discrepancy Theory" (September 7, 2002), https://arxiv.org/abs/math/0209078.

7. Michael Held and Richard M. Karp, "The Traveling-Salesman Problem and Minimum Spanning Trees," *Operations Research* (December 1, 1970): 1138–1162, http://pubsonline.informs.org/doi/pdf/10.1287/opre.18.6.1138.

8. Nima Anari and Shayan Oveis Gharan, "The Kadison–Singer Problem for Strongly Rayleigh Measures and Applications to Asymmetric TSP" (July 22, 2015), https://arxiv.org/pdf/1412.1143v2.pdf.

Mathematicians Tame Rogue Waves, Lighting Up Future of LEDs

1. Marcel Filoche and Svitlana Mayboroda, "Universal Mechanism for Anderson and Weak Localization," *Proceedings of the National Academy of Sciences* 109, no. 37 (September 2012): 14761–14766, http://www.pnas.org/content/109/37/14761.full.pdf.

Pentagon Tiling Proof Solves Century-Old Math Problem

1. R. B. Kershner, "On Paving the Plane," *American Mathematical Monthly* 75, no. 8 (October 1968): 839–844, http://www.jhuapl.edu/techdigest/views/pdfs/V08_N6 _1969/V8_N6_1969_Kershner.pdf.

2. Casey Mann, Jennifer McLoud-Mann and David Von Derau, "Convex Pentagons That Admit I-Block Transitive Tilings" (October 5, 2015), https://arxiv.org/abs/1510 .01186.

3. Michaël Rao, "Exhaustive Search of Convex Pentagons Which Tile the Plane" (August 1, 2017), https://perso.ens-lyon.fr/michael.rao/publi/penta.pdf.

4. Emmanuel Jeandel and Michaël Rao, "An Aperiodic Set of 11 Wang Tiles" (June 22, 2015), https://arxiv.org/abs/1506.06492.

Simple Set Game Proof Stuns Mathematicians

1. R. Hill, "On Pellegrino's 20-Caps in $S_{4, 3}$," *North-Holland Mathematics Studies* 78 (1983): 433–447, http://www.sciencedirect.com/science/article/pii/S0304020808 73322X.

2. Roy Mechulam, "On Subsets of Finite Abelian Groups with No 3-Term Arithmetic Progressions," *Journal of Combinatorial Theory, Series A* 71, no. 1 (July 1995): 168–172, http://www.sciencedirect.com/science/article/pii/0097316595900241; Michael Bateman and Nets Hawk Katz, "New Bounds on Cap Sets," *Journal of the American Mathematical Society* 25 (2012): 585–613, http://www.ams.org/journals/jams/2012-25-02/ S0894-0347-2011-00725-X/.

3. Ernie Croot, Vsevolod Lev and Peter Pach, "Progression-Free Sets in Z_4^n Are Exponentially Small" (May 21, 2016), https://arxiv.org/abs/1605.01506.

4. Jordan S. Ellenberg and Dion Gijswijt, "On Large Subsets of $\mathbb{F}_q^n$ with No Three-Term Arithmetic Progression," *Annals of Mathematics* 185, no. 1 (2017): 339–343, http://annals.math.princeton.edu/2017/185-1/p08.

5. Jonah Blasiak et al., "On Cap Sets and the Group-Theoretic Approach to Matrix Multiplication," *Discrete Analysis* 2017, no. 3, https://arxiv.org/abs/1605.06702.

A Magical Answer to an 80-Year-Old Puzzle

1. Boris Konev and Alexei Lisitsa, "A SAT Attack on the Erdős Discrepancy Conjecture" (February 17, 2014), https://arxiv.org/pdf/1402.2184.pdf.

2. Terence Tao, "The Erdős Discrepancy Problem" (January 13, 2017), https://arxiv.org/pdf/1509.05363v1.pdf.

3. Kaisa Matomäki and Maksym Radziwiłł, "Multiplicative Functions in Short Intervals" (October 15, 2017), https://arxiv.org/pdf/1501.04585v2.pdf.

Sphere Packing Solved in Higher Dimensions

1. Thomas C. Hales, "A Proof of the Kepler Conjecture," *Annals of Mathematics* 162, no. 3 (2005): 1065–1185, http://annals.math.princeton.edu/2005/162-3/p01.

2. Maryna Viazovska, "The Sphere Packing Problem in Dimension 8" (April 4, 2017), https://arxiv.org/abs/1603.04246.

3. John Leech, "Some Sphere Packings in Higher Space," *Canadian Journal of Mathematics* 16, (January 1964): 657–682, https://cms.math.ca/10.4153/CJM-1964-065-1.

4. John Leech, "Notes on Sphere Packings," *Canadian Journal of Mathematics* 19 (1967): 251–267, https://cms.math.ca/10.4153/CJM-1967-017-0; J. H. Conway, "A Perfect Group of Order 8,315,553,613,086,720,000 and the Sporadic Simple Groups," *Proceedings of the National Academy of Sciences* 61, no. 2 (October 1968): 398–400, http://www.pnas.org/content/61/2/398; J. H. Conway, "A Group of Order 8,315,553,613,086,720,000," *Bulletin of the London Mathematical Society* 1, no. 1 (March 1, 1969): 79–88, https://doi.org/10.1112/blms/1.1.79.

5. Henry Cohn and Noam Elkies, "New Upper Bounds on Sphere Packings I," *Annals of Mathematics* 157 (2003): 689–714, http://annals.math.princeton.edu/wp-content/uploads/annals-v157-n2-p09.pdf.

6. Henry Cohn and Abhinav Kumar, "Optimality and Uniqueness of the Leech Lattice among Lattices," *Annals of Mathematics* 170, no. 3 (2009): 1003–1050, http://annals.math.princeton.edu/2009/170-3/p01.

7. Henry Cohn et al., "The Sphere Packing Problem in Dimension 24," *Annals of Mathematics* 185 (2017): 1017–1033, https://arxiv.org/abs/1603.06518.

A Tenacious Explorer of Abstract Surfaces

1. Alex Eskin and Maryam Mirzakhani, "Invariant and Stationary Measures for the SL(2,R) Action on Moduli Space" (November 26, 2017), httpx://arxiv.org/pdf/1302.3320.pdf.

2. Maryam Mirzakhani, "Growth of the Number of Simple Closed Geodesics on Hyperbolic Surfaces," *Annals of Mathematics* 168, no. 1 (2008): 97–125, http://annals .math.princeton.edu/2008/168-1/p03; "Simple Geodesics and Weil-Petersson Volumes of Moduli Spaces of Bordered Riemann Surfaces," *Inventiones Mathematicae* 167, no. 1 (January 2007): 179–222, httpx://link.springer.com/article/10.1007/s00222 -006-0013-2; "Weil-Petersson Volumes and Intersection Theory on the Moduli Space of Curves," *Journal of the American Mathematics Society* 20 (2007): 1–23, http://www .ams.org/journals/jams/2007-20-01/S0894-0347-06-00526-1/home.html.

3. Alex Eskin, Maryam Mirzakhani and Amir Mohammadi, "Isolation, Equidistribution and Orbit Closures for the *SL*(2,R) Action on Moduli Space" (March 2, 2015), https://arxiv.org/pdf/1305.3015.pdf.

4. Alex Eskin and Maryam Mirzakhani, "Invariant and Stationary Measures for the *SL*(2, R) Action on Moduli Space" (November 26, 2017), https://arxiv.org/pdf/1302 .3320.pdf.

5. Samuel Lelièvre, Thierry Monteil and Barak Weiss, "Everything Is Illuminated," *Geometry & Topolology* 20 (2016): 1737–1762, https://arxiv.org/pdf/1407 .2975.pdf.

6. Anton Zorich, "Flat Surfaces," in *Frontiers in Number Theory, Physics, and Geometry I*, ed. P. Cartier et al. (Berlin: Springer, 2006), 437–583, https://arxiv.org/pdf/math /0609392.pdf.

A "Rebel" Without a Ph.D.

1. William H. Press and Freeman J. Dyson, "Iterated Prisoner's Dilemma Contains Strategies That Dominate Any Evolutionary Opponent," *Proceedings of the National Academy of Sciences* 109, no. 26 (June 2012): 10409–10413, http://www.pnas.org/con tent/109/26/10409.full?sid=170efdfd-ac48-4ea2-9851-064e11184b81.

2. Freeman Dyson, *The Scientist as Rebel* (New York: New York Review Books, 2006), https://books.google.com/books?id=dfe_s_tK080C&printsec=frontcover&source =gbs_ge_summary_r&hl=en.

3. William H. Press, "Bandit Solutions Provide Unified Ethical Models for Randomized Clinical Trials and Comparative Effectiveness Research," *Proceedings of the National Academy of Sciences* 106, no. 52 (December 2009): 22387–22392, http:// www.pnas.org/content/106/52/22387.

A Brazilian Wunderkind Who Calms Chaos

1. Mikhail Lyubich, "Almost Every Real Quadratic Map Is Either Regular or Stochastic" (July 15, 1997), https://arxiv.org/abs/math/9707224.

2. P. Coullet and C. Tresser, "Itérations *d'endomorphismes et groupe* de renormalization," *Journal de Physique Colloques* 39 (1978): C5–28, https://hal.archives-ouvertes.fr/docs/00/21/74/80/PDF/ajp-jphyscol197839C513.pdf.

3. Mikhail Lyubich, "Forty Years of Unimodal Dynamics: On the Occasion of Artur Avila Winning the Brin Prize," *Journal of Modern Dynamics* 6, no. 2 (2012), 183–203, http://www.math.stonybrook.edu/~mlyubich/papers/Brin-prize.pdf.

4. Artur Avila and Jairo Bochi, "A Generic C^1 map Has No Absolutely Continuous Invariant Probability Measure," *Nonlinearity* 19, no. 11 (October 18, 2006), http://www.mat.uc.cl/~jairo.bochi/docs/acim.pdf.

5. Artur Avila and Svetlana Jitomirskaya, "The Ten Martini Problem," *Annals of Mathematics* 170, no. 1 (2009): 303–342, http://annals.math.princeton.edu/2009/170-1/p08.

6. Artur Avila, Sylvain Crovisier and Amie Wilkinson, "Diffeomorphisms with Positive Metric Entropy," *Publications mathématiques de l'IHÉS* 124 (2016), 319–347, http://www.math.uchicago.edu/~wilkinso/papers/acw-august2014.pdf.

The Musical, Magical Number Theorist

1. Manjul Bhargava, "Higher Composition Laws I: A New View on Gauss Composition, and Quadratic Generalizations," *Annals of Mathematics* 159, no. 1 (January 2004): 217–250, http://www.jstor.org/stable/3597249; "Higher Composition Laws II: On Cubic Analogues of Gauss Composition," *Annals of Mathematics* 159, no. 2 (March 2004): 865–886, http://www.jstor.org/stable/3597310.

2. Manjul Bhargava and Arul Shankar, "The Average Size of the 5-Selmer Group of Elliptic Curves Is 6, and the Average Rank Is Less Than 1" (December 31, 2013), https://arxiv.org/pdf/1312.7859.pdf.

3. Manjul Bhargava and Christopher Skinner, "A Positive Proportion of Elliptic Curves Over Q Have Rank One" (January 3, 2014), https://arxiv.org/pdf/1401.0233.pdf.

4. Manjul Bhargava, Christopher Skinner and Wei Zhang, "A Majority of Elliptic Curves over Q Satisfy the Birch and Swinnerton-Dyer Conjecture" (July 17, 2014), https://arxiv.org/pdf/1407.1826.pdf.

The Oracle of Arithmetic

1. Peter Scholze, "The Local Langlands Correspondence for GL_n over *p*-adic Fields" (October 10, 2010), https://arxiv.org/abs/1010.1540.

2. Peter Scholze, "Perfectoid Spaces" (November 21, 2011), https://arxiv.org/abs/1111.4914.

3. Peter Scholze, "On Torsion in the Cohomology of Locally Symmetric Varieties" (June 2, 2015), https://arxiv.org/abs/1306.2070.

In Noisy Equations, One Who Heard Music

1. Martin Hairer, "A Theory of Regularity Structures" (February 15, 2014), https://arxiv.org/abs/1303.5113.

2. Martin Hairer and Jonathan C. Mattingly, "Ergodicity of the 2D Navier–Stokes Equations with Degenerate Stochastic Forcing" (April 26, 2007), https://arxiv.org/pdf/math.PR/0406087.pdf.

3. Mehran Kardar, Giorgio Parisi and Yi-Cheng Zhang, "Dynamic Scaling of Growing Interfaces," *Physical Review Letters* 56, no. 9 (March 3, 1986), https://journals.aps.org/prl/abstract/10.1103/PhysRevLett.56.889.

4. Martin Hairer, "Solving the KPZ Equation" (July 26, 2012), https://arxiv.org/pdf/1109.6811v3.pdf.

Michael Atiyah's Imaginative State of Mind

1. Roger Penrose, "Palatial Twistor Theory and the Twistor Googly Problem," *Philosophical Transactions of the Royal Society A* (June 29, 2015), http://rsta.royalsocietypublishing.org/content/373/2047/20140237.

2. M. F. Atiyah and I. M. Singer, "The Index of Elliptic Operators on Compact Manifolds," *Bulletin of the American Mathematical Society* 69 (1963): 422–433, http://www.ams.org/journals/bull/1963-69-03/S0002-9904-1963-10957-X/home.html.

3. Semir Zeki et al., "The Experience of Mathematical Beauty and Its Neural Correlates," *Frontiers of Human Neuroscience* 13 (February 13, 2014), http://www.frontiersin.org/article/10.3389/fnhum.2014.00068/full.

4. A. Einstein, "The Field Equations of Gravitation," http://einsteinpapers.press.princeton.edu/vol6-trans/129.

Landmark Algorithm Breaks 30-Year Impasse

1. László Babai, "Graph Isomorphism in Quasipolynomial Time" (January 19, 2016), https://arxiv.org/abs/1512.03547v1.

2. László Babai and Eugene M. Luks, "Canonical Labeling of Graphs," *STOC '83 Proceedings of the Fifteenth Annual ACM Symposium on Theory of Computing* (1983): 171–183, https://dl.acm.org/citation.cfm?id=808746.

A Grand Vision for the Impossible

1. Subhash Khot, "On the Power of Unique 2-Prover 1-Round Games," *STOC '02 Proceedings of the Thirty-Fourth Annual ACM Symposium on Theory of Computing* (2002): 767–775, https://dl.acm.org/citation.cfm?id=510017.

2. Sanjeev Arora et al., "Proof Verification and the Hardness of Approximation Problems," *Journal of the ACM* 45, no. 3 (May 1998): 501–555, https://dl.acm.org/citation .cfm?doid=278298.278306.

3. Johan Håstad, "Some Optimal Results," *Journal of the ACM* 48, no. 4 (July 2001): 798–859, https://dl.acm.org/citation.cfm?id=502098.

4. Khot, "On the Power of Unique 2-Prover 1-Round Games."

5. Subhash Khot and Oded Regev, "Vertex Cover Might Be Hard to Approximate to within 2-ε," *Journal of Computer and System Sciences* 74, no. 3 (May 2008): 335–349, https://dl.acm.org/citation.cfm?id=1332256.

6. Subhash Khot et al., "Optimal Inapproximability Results for MAX-CUT and Other 2-Variable CSPs?" *SIAM Journal on Computing* 37, no. 1 (April 2007): 319–357, https:// dl.acm.org/citation.cfm?id=1328735.

7. Prasad Raghavendra, "Optimal Algorithms and Inapproximability Results for Every CSP?" *STOC '08 Proceedings of the Fortieth Annual ACM Symposium on Theory of Computing* (May 17–20, 2008): 245–254, https://people.eecs.berkeley.edu/~prasad /Files/extabstract.pdf.

8. Guy Kindler et al., "Spherical Cubes: Optimal Foams from Computational Hardness Amplification," *Communications of the ACM* 55, no. 10 (October 2012): 90–97, https://dl.acm.org/citation.cfm?id=2347757; Subhash A. Khot and Nisheeth K. Vishnoi, "The Unique Games Conjecture, Integrality Gap for Cut Problems and Embeddability of Negative Type Metrics into ℓ_1" (May 20, 2013), https://cs.nyu.edu /~khot/papers/gl-journal-ver1.pdf; Elchanan Mossel, Ryan O'Donnell and Krzysztof Oleszkiewicz, "Noise Stability of Functions with Low Influences: Invariance and Optimality," *Annals of Mathematics* 171, no. 1 (2010): 295–341, http://annals.math .princeton.edu/2010/171-1/p05.

To Settle Infinity Dispute, a New Law of Logic

1. W. Hugh Woodin, "Strong Axioms of Infinity and the Search for *V*," *Proceedings of the International Congress of Mathematicians* (2010), http://logic.harvard.edu /EFI_Woodin_StrongAxiomsOfInfinity.pdf.

2. David Asperó, Paul Larson and Justin Tatch Moore, "Forcing Axioms and the Continuum Hypothesis," *Acta Mathematica* 210, no. 1 (2013): 1–29, http://www .users.miamioh.edu/larsonpb/Pi2_CH.pdf.

3. Joseph Warren Dauben, *Georg Cantor: His Mathematics and Philosophy of the Infinite* (Princeton, N.J.: Princeton University Press, 1990), https://books.google.com/ books/about/Georg_Cantor.html?id=-cpFeTPJXDIC&hl=en.

4. W. Hugh Woodin, "Suitable Extender Models I," *Journal of Mathematical Logic* 10, no. 101 (2010), http://www.worldscientific.com/doi/abs/10.1142/S021906131000095X ?journalCode=jml.

5. M. Foreman, M. Magidor and S. Shelah, "Martin's Maximum, Saturated Ideals and Non-Regular Ultrafilters," *Annals of Mathematics* 127, no. 1 (January 1988): 1–47, http://www.jstor.org/stable/10.2307/1971415.

6. Stevo Todorčević and Peter Koellner, "The Power-Set of ω_1 and the Continuum Problem," http://logic.harvard.edu/Todorčević_Structure4.pdf.

Mathematicians Bridge Finite–Infinite Divide

1. Ludovic Patey and Keita Yokoyama, "The Proof-Theoretic Strength of Ramsey's Theorem for Pairs and Two Colors" (April 26, 2016), https://arxiv.org/abs/1601 .00050.

2. W. W. Tait, "Primitive Recursive Arithmetic and its Role in the Foundations of Arithmetic: Historical and Philosophical Reflections in Honor of Per Martin-Löf on the Occasion of His Retirement," http://home.uchicago.edu/~wwtx/PRA2.pdf.

Mathematicians Measure Infinities and Find They're Equal

1. M. Malliaris and S. Shelah, "Cofinality Spectrum Theorems in Model Theory, Set Theory, and General Topology," *Journal of the American Mathematical Society* 29 (2016): 237–297, http://www.ams.org/journals/jams/2016-29-01/S0894-0347-2015 -00830-X/home.html.

2. Justin Tatch Moore, "Model Theory and the Cardinal Numbers p and t," *Proceedings of the National Academy of Sciences* 110, no. 33 (August 2013): 13238–13239, http://www .pnas.org/content/110/33/13238.full.pdf.

A Life Inspired by an Unexpected Genius

1. Ken Ono and Amir D. Aczel, *My Search for Ramanujan: How I Learned to Count* (Cham, Switzerland: Springer, 2016), http://www.springer.com/us/book/9783319255668.

INDEX